It's another great book from CGP...

Biology exams can seem daunting — especially if you're not sure what to expect.
But they're less stressful if you've done plenty of realistic practice in advance.

Happily, this book (which includes a **free** Online Edition)
is packed with exam-style questions for every topic. It even includes
two complete practice exams to fully prepare you for the real thing.

How to get your free online edition

Want to read this book on your computer or tablet?
Just go to **cgpbooks.co.uk/extras** and enter this code...

0302 2031 3689 8549

By the way, this code only works for one person. If somebody else has used
this book before you, they might have already claimed the online edition.

CGP — still the best! ☺

Our sole aim here at CGP is to produce the highest quality books —
carefully written, immaculately presented and dangerously close to being funny.

Then we work our socks off to get them out to you
— at the cheapest possible prices.

Contents

☑ Use the tick boxes to check off the topics you've completed.

Section Six — Plants

Section Seven — Variation and Genetics

Section Eight — Adaptation and Environment

Section Nine — Environment, Evolution and Energy Transfer

Practice Papers

How to get answers for the Practice Papers: Your free Online Edition of this book includes all the answers for Practice Papers 1 & 2. (Just flick back to the first page to find out how to get hold of your Online Edition.)

Published by CGP

Editors:
Katie Braid, Christopher Lindle, Rachael Marshall, Rachael Rogers, Jo Sharrock,
Hayley Thompson, Rachel Ward.

With thanks to Janet Cruse-Sawyer and Karen Wells for the proofreading.
With thanks to Philip Rushworth for the reviewing.
With thanks to Laura Jakubowski for the copyright research.

ISBN: 978 1 84762 447 5

Data used to construct the graph on page 19 reproduced with kind permission from
Oxford University Press. P. Reynolds, et al. Active Smoking, Household Passive Smoking,
and Breast Cancer: Evidence From the California Teachers Study. JNCI 2004; 96(1):29-37.

Data about kidney transplant waiting lists on page 45 from http://www.organdonation.nhs.uk/
statistics/transplant_activity_report/archive_activity_reports/pdf/ukt/activity_report_2011_12.pdf
© NHS Blood and Transplant

Data used to construct the table about treatments for kidney failure on page 45 from
http://www.organdonation.nhs.uk/newsroom/fact_sheets/cost_effectiveness_of_
transplantation.asp © NHS Blood and Transplant

With thanks to the IPCC for permission to reproduce the graph on page 88. Climate Change 2007:
The Physical Science Basis. Working Group I Contribution to the Fourth Assessment Report of the
Intergovernmental Panel on Climate Change.

Data used to construct the graph on page 89 from International Journal of Biometeorology February
1999, Volume 42, Issue 3, pp 134-138. Phenology and the changing pattern of bird migration in
Britain by T. H. Sparks. Used with permission from Springer. www.springer.com

DDT diagram on page 95 from Biological Science Combined Volume Hardback,
1990, Soper, Green, Stout, Taylor. Cambridge University Press.

Data on page 98 relating to eggs in the UK, source: www.defra.gov.uk © Crown Copyright
reproduced under the terms of the Open Government licence http://www.nationalarchives.gov.uk/
doc/open-government-licence/

Every effort has been made to locate copyright holders and obtain permission to reproduce sources.
For those sources where it has been difficult to trace the originator of the work, we would be
grateful for information. If any copyright holder would like us to make an amendment to the
acknowledgements, please notify us and we will gladly update the book at the next reprint.
Thank you.

Clipart from Corel®
Printed by Elanders Ltd, Newcastle upon Tyne

Based on the classic CGP style created by Richard Parsons.

Text, design, layout and original illustrations © Coordination Group Publications Ltd. (CGP) 2013
All rights reserved.

How to Use This Book

- Hold the book <u>upright</u>, approximately <u>50 cm</u> from your face, ensuring that the text looks like <u>this</u>, not sıɥʇ. Alternatively, place the book on a <u>horizontal</u> surface (e.g. a table or desk) and sit adjacent to the book, at a distance which doesn't make the text too small to read.

- In case of emergency, press the two halves of the book together <u>firmly</u> in order to close.

- Before attempting to use this book, familiarise yourself with the following <u>safety information</u>:

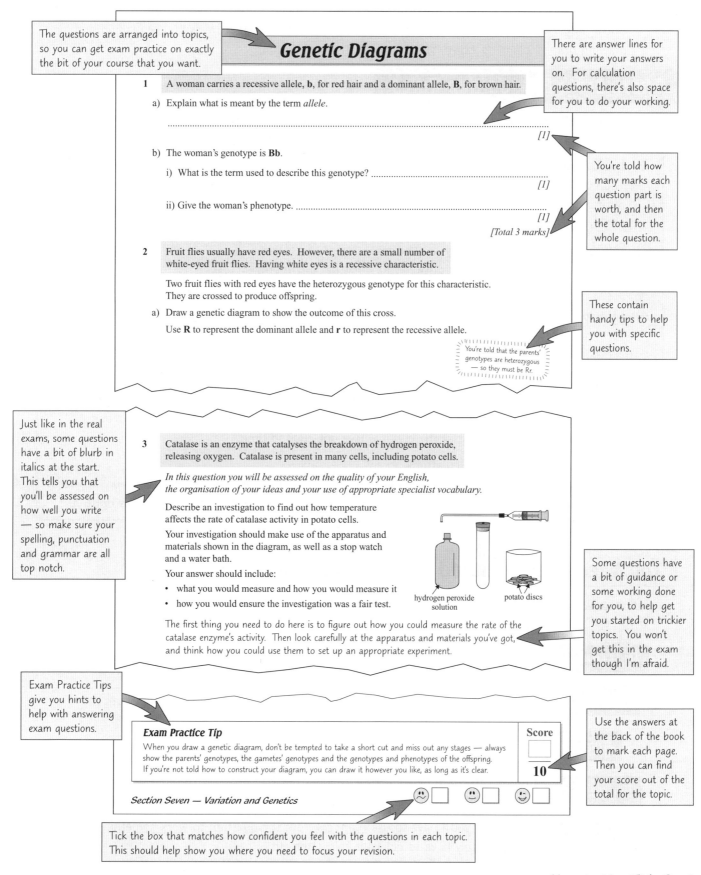

The questions are arranged into topics, so you can get exam practice on exactly the bit of your course that you want.

Genetic Diagrams

There are answer lines for you to write your answers on. For calculation questions, there's also space for you to do your working.

1   A woman carries a recessive allele, **b**, for red hair and a dominant allele, **B**, for brown hair.

a) Explain what is meant by the term *allele*.

   ....
                                                                                              [1]

b) The woman's genotype is **Bb**.

   i) What is the term used to describe this genotype? ....
                                                                                              [1]

   ii) Give the woman's phenotype. ....
                                                                                              [1]
                                                                              [Total 3 marks]

You're told how many marks each question part is worth, and then the total for the whole question.

2   Fruit flies usually have red eyes. However, there are a small number of white-eyed fruit flies. Having white eyes is a recessive characteristic.

Two fruit flies with red eyes have the heterozygous genotype for this characteristic. They are crossed to produce offspring.

a) Draw a genetic diagram to show the outcome of this cross.

   Use **R** to represent the dominant allele and **r** to represent the recessive allele.

These contain handy tips to help you with specific questions.

You're told that the parents' genotypes are heterozygous — so they must be Rr.

Just like in the real exams, some questions have a bit of blurb in italics at the start. This tells you that you'll be assessed on how well you write — so make sure your spelling, punctuation and grammar are all top notch.

3   Catalase is an enzyme that catalyses the breakdown of hydrogen peroxide, releasing oxygen. Catalase is present in many cells, including potato cells.

In this question you will be assessed on the quality of your English, the organisation of your ideas and your use of appropriate specialist vocabulary.

Describe an investigation to find out how temperature affects the rate of catalase activity in potato cells.

Your investigation should make use of the apparatus and materials shown in the diagram, as well as a stop watch and a water bath.

Your answer should include:

- what you would measure and how you would measure it
- how you would ensure the investigation was a fair test

hydrogen peroxide solution   potato discs

The first thing you need to do here is to figure out how you could measure the rate of the catalase enzyme's activity. Then look carefully at the apparatus and materials you've got, and think how you could use them to set up an appropriate experiment.

Some questions have a bit of guidance or some working done for you, to help get you started on trickier topics. You won't get this in the exam though I'm afraid.

Exam Practice Tips give you hints to help with answering exam questions.

Exam Practice Tip

When you draw a genetic diagram, don't be tempted to take a short cut and miss out any stages — always show the parents' genotypes, the gametes' genotypes and the genotypes and phenotypes of the offspring. If you're not told how to construct your diagram, you can draw it however you like, as long as it's clear.

Score

10

Section Seven — Variation and Genetics

Use the answers at the back of the book to mark each page. Then you can find your score out of the total for the topic.

Tick the box that matches how confident you feel with the questions in each topic. This should help show you where you need to focus your revision.

Exam Tips

AQA Certificate Exam Stuff

1) You have to do two exams for the AQA Level 1/2 Certificate in Biology — Paper 1 and Paper 2 (ingenious).

2) Both exams are 1½ hours long, and worth 90 marks.

3) Both papers test your knowledge and understanding of Biology. No surprises there.
   But in Paper 2, there's more of a focus on experimental and investigative skills, like reading and drawing graphs, planning experiments, or evaluating conclusions.

There are a Few Golden Rules

1) **Always, always, always make sure you read the question properly.**
   For example, if the question asks you to give your answer in mm, don't give it in cm.

2) **Look at the number of marks a question is worth.**
   The number of marks gives you a pretty good clue of how much to write.
   So if a question is worth four marks, make sure you write four decent points. And there's no point writing an essay for a question that's only worth one mark — it's just a waste of your time.

3) **Write your answers as clearly as you can.**
   If the examiner can't read your answer you won't get any marks, even if it's right.

4) **Use specialist vocabulary.**
   You know the words I mean — the silly sciencey ones, like extremophile and differentiation. Examiners love them.

> Obeying these Golden Rules will help you get as many marks as you can in the exam — but they're no use if you haven't learnt the stuff in the first place. So make sure you revise well and do as many practice questions as you can.

5) **Pay attention to the time.**
   The amount of time you've got for each paper means you should spend about a minute per mark.
   So if you're totally, hopelessly stuck on a question, just leave it and move on to the next one.
   You can always go back to it at the end if you've got enough time.

6) **Show each step in your calculations.**
   You're less likely to make a mistake if you write things out in steps. And even if your final answer's wrong, you'll probably pick up some marks if the examiner can see that your method is right.

You Need to Understand the Command Words

Command words are the words in a question that tell you what to do.
If you don't know what they mean, you might not be able to answer the questions properly.

Describe... This means you need to recall facts or write about what something is like.

Explain... You have to give reasons for something or say why or how something happens.

Give... This means the same thing as 'Name...' or 'State...'.
You usually just have to give a short definition or an example of something.

Suggest... You need to use your knowledge to work out the answer. It'll often be something you haven't been taught, but you should be able to use what you know to figure it out.

Calculate... This means you'll have to use numbers from the question to work something out.
You'll probably have to get your calculator out.

Cells

1 Animal cells contain a number of different parts, all of which have specific functions.

State the function of the following cell parts.

a) Cell membrane: ...*Controls what goes into and out of the cell*...........

[1]

b) Ribosomes: ....*produce proteins*...

[1]

c) Mitochondria: ...*where chemical reactions take place in a cell and provide energy*

[1]

[Total 3 marks]

2 Plant and animal cells have similarities and differences.

a) A plant cell is shown below.

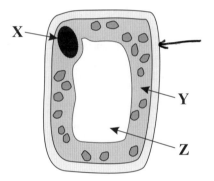

X →

Y

Z

i) Name the parts of the cell labelled **X**, **Y** and **Z**.

X ...*Nucleus*..

Y ..*Cytoplasm*..

Z ..*Permanent Vacuole*...............................

[3]

ii) Add an arrow to the diagram to show where cellulose would be found.

[1]

b) Give **one** similarity and **one** difference between the structure of a human cell and the structure of a plant cell.

Similarity: ...*They both have a nucleus*....................................

Difference: ..*Only plant cells have a cell wall or chloroplasts*...

[2]

[Total 6 marks]

3 The diagrams on the right show a yeast cell and a bacterial cell.

A

nucleus
containing
genetic material   cell membrane

cell
wall

B

cell
membrane   genetic
material

cell
wall

plasmid

Give **two** reasons why cell **B** must be the bacterial cell and not the yeast cell.

..*Because the genetic material is found in the*.......

..*cytoplasm of bacterial cells, not the nucleus*...........

[Total 2 marks]

Score:

11

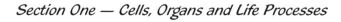

Specialised Cells

1 Sperm and egg cells are specially adapted for their roles in reproduction. The sperm cell's role is to swim through the female reproductive system and carry the male DNA to the egg cell. The egg cell contains the female DNA and supports the initial stages of the developing embryo.

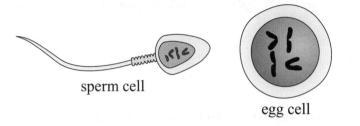

sperm cell

egg cell   Not to scale.

a) i) Using the diagrams above, suggest **one** feature of a sperm cell that makes it well adapted for its function. Explain your answer.

...It..has..a..long..tail..to..swim..to..the..egg..cell..

....

[2]

ii) Sperm cells contain large numbers of mitochondria. Suggest why.

...To....provide...energy...to..swim...to..the...egg..cell..................................

[1]

b) Egg cells contain large food reserves. Suggest why these are necessary.

...To...feed...the..embryo..

[1]

c) Explain why both the sperm cell and the egg cell contain a nucleus.

...To.....carry...the....female...and..male...genetic...~~information~~...................

.material.../..DNA...

[1]

[Total 5 marks]

2 White blood cells help to defend the body against disease. One way they do this is to produce proteins called antibodies, which destroy disease-causing bacteria in the bloodstream.

Comment on the quantity of ribosomes that you might expect white blood cells to contain. Explain your answer.

...You...would..expect...white...blood....cell..to..have..a..large..number....

.of....ribosomes....to...allow...them...to...produce...a..large..number...of......

.antibodies.../.proteins..

[Total 2 marks]

Score: ☐

7

☹ ☐   ☺ ☐   🙂 ☐

Tissues, Organs and Organ Systems

1   The human digestive system is shown in the diagram below.

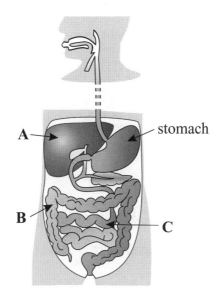

a)   Name the organs labelled **A** - **C** on the diagram and describe their function.

   i)   Organ **A**

       Name: ...*Liver*..

       Function: ...*produces bile*..

                                                                                              [2]

   ii)   Organ **B**

       Name: ...*large intestine*..

       Function: ....

                                                                                              [2]

   iii) Organ **C**

       Name: ....*Small intestine*..

       Function: ....

                                                                                              [2]

b)   Glandular tissue in the stomach produces digestive juices to digest food.

   i)   Name **two** other glands within the digestive system that also have this function.

       1. ...*Salivary glands*...

       2. ..*pancreas*...

                                                                                              [2]

   ii)   Name **one** other type of tissue found in the stomach and state its function.

       muscular tissue - allows food to be churned in the stomach
                                                                                              [2]

                                                                                    [Total 10 marks]

Section One — Cells, Organs and Life Processes

2 The diagram below shows the human urinary system. Its main function is to produce urine and remove it from the body. Urine is made in the kidneys, before travelling down tubes called the ureters and then into the bladder where it is stored.

kidneys

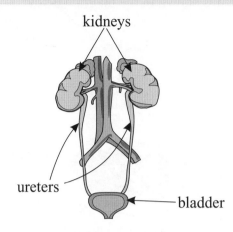

ureters

bladder

a) The urinary system is an *organ system*. What is meant by the term *organ system*?

...A group of different organs that work together to perform a job..................

[1]

b) The adrenal medullas are located on top of the kidneys.
They are responsible for secreting the hormone adrenaline in response to stress.

State **one** type of tissue the adrenal medulla must contain. Explain your answer.

..Glandular tissue, because the adrenal medulla secretes hormones, which is the function of glandular tissue..

[2]

c) The bladder and the ureters contain both epithelial and muscular tissue.

i) Suggest and explain why the ureters contain muscular tissue.

...So that the urine is able to travel down the ureters into the bladder..................

[2]

ii) Explain why the bladder and the ureters can be classed as organs.

...Because they are made up of tissues..................

[1]

d) During the development of the urinary system, its cells differentiate.
Explain why they must do this.

....To become specialised at certain functions..................

[1]

[Total 7 marks]

Score: ☐

17

😕 ☐   😐 ☐   🙂 ☐

Plant Tissues and Organs

1   Plants are made up of cells, tissues, organs and organ systems.

a) Name **three** organs found in a plant.

...roots, stems, leaves..

[3]

Plants contain several different tissues.

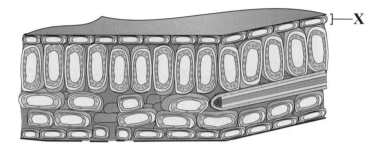

]—X

b) i) Name the tissue labelled **X** on the diagram above.

...epidermal tissue...

[1]

ii) Name the **two** types of mesophyll tissue present in leaves and describe the function of each.

1. ...spongy mesophyll tissue - contains air spaces to allow gases to diffuse in and out of cells..............................

2. ...palisade mesophyll tissue - where most photosynthesis happens.................................

[4]

[Total 8 marks]

2   The diagram below shows a cross-section of a dicotyledonous plant stem.

a) i) Name the tissues labelled on the diagram.

A ..Phloem..............................

B ..xylem..............................

[2]

ii) What is the function of these tissues?

....

[1]

b) Describe **one** way in which a cross-section of a root would look different from the cross-section of a stem.

....

....

[1]

Score:

[Total 4 marks]

12

Diffusion

1 The diagram on the right shows a cell and the surrounding tissue fluid. Oxygen moves in and out of the cell by diffusion.

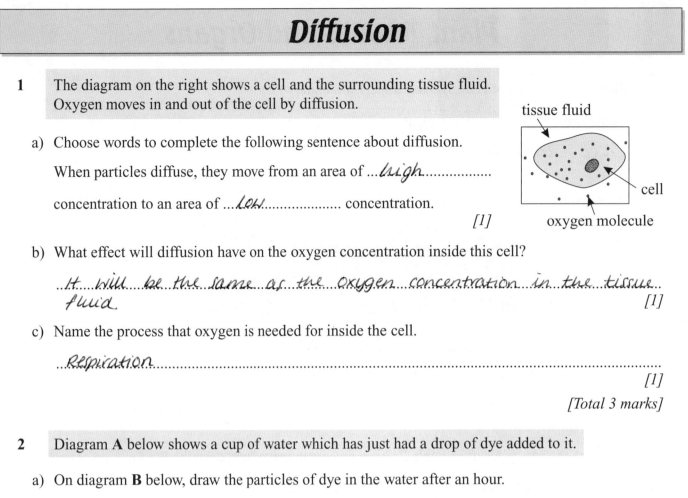

tissue fluid

cell

oxygen molecule

a) Choose words to complete the following sentence about diffusion.

When particles diffuse, they move from an area of ...*high*...

concentration to an area of ...*low*... concentration.

[1]

b) What effect will diffusion have on the oxygen concentration inside this cell?

...*It will be the same as the oxygen concentration in the tissue fluid*...

[1]

c) Name the process that oxygen is needed for inside the cell.

...*Respiration*...

[1]

[Total 3 marks]

2 Diagram **A** below shows a cup of water which has just had a drop of dye added to it.

a) On diagram **B** below, draw the particles of dye in the water after an hour.

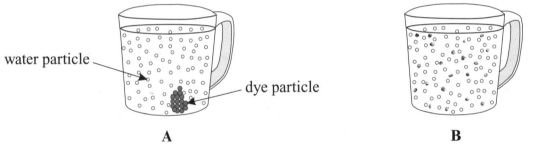

water particle

dye particle

A

B

[1]

b) Predict how the rate of diffusion of the dye would change if a large drop of dye were used instead of a small drop of dye. Explain your answer.

...*The rate of diffusion would be faster*...

...

[2]

[Total 3 marks]

3 A student was given three solutions labelled X, Y and Z. He placed equal amounts of each solution inside a bag designed to act like a cell membrane. He then put the bag inside a beaker of water. After 24 hours, the water outside the bag contained particles X and Y, but not Z.

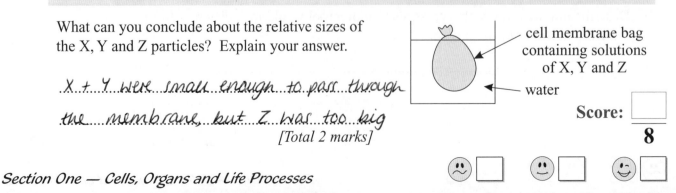

What can you conclude about the relative sizes of the X, Y and Z particles? Explain your answer.

cell membrane bag containing solutions of X, Y and Z

water

...*X + Y were small enough to pass through the membrane, but Z was too big*...

[Total 2 marks]

Score:

8

Section One — Cells, Organs and Life Processes

Osmosis

1   The diagram below shows some body cells bathed in tissue fluid. A blood vessel flows close
     to the cells, providing water. The cells shown have a low concentration of water inside them.

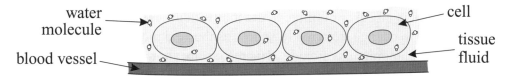

water
molecule → ← cell

blood vessel → ← tissue fluid

   a) i) Water moves into and out of cells by *osmosis*. What is meant by the term *osmosis*?

   ...Water...molecules...move...through...a...partially...permeable...membrane
   from...a...region...of...high...water...concentration...to...a...region...of...low...
   water concentration
                                                                              [2]

       ii) Use osmosis to explain whether the net movement of water in the diagram would be into
           or out of the body cells.

   ...Water...molecules...would...move...into...the...cell...because...there...is...a
   lower...concentration...inside...the...cell...(~~hyper~~ hypotonic)...
                                                                              [2]

   b) Is the tissue fluid in the diagram a hypertonic, hypotonic or isotonic solution?
       Explain your answer.

   ...Hypotonic...because...the...concentration...of...water...is...higher...
   outside...the...cell...
                                                                              [2]

   c) Suggest why osmosis appears to stop after a while.

   ...When...there...is...an...equilibrium...
                                                                              [1]

                                                                    [Total 7 marks]

2   Plant cells look different depending on how much water they contain.

   a) Use the words in the box below to describe the states of the following cells.

plasmolysed	turgid	normal	flaccid

   A ...Normal...   B ...turgid...   C ...flaccid...   D ...plasmolysed...
                                                                              [2]

   b) Explain why plants start to wilt if they don't have enough water.

   ....

   ....

   ....
                                                            [2]   Score:

                                                                              11
                                                     [Total 4 marks]

Diffusion and Osmosis Experiments

1   A student made up some gelatine containing cresol red solution and ammonium hydroxide. Cresol red solution is a pH indicator that is red in alkaline solutions and yellow in acidic solutions. The student cut the gelatine into cubes of different sizes, and placed the cubes in a beaker of dilute hydrochloric acid. He measured how long it took for the cubes to change from red to yellow as the acid moved into the gelatine and neutralised the ammonium hydroxide. His results are shown in the table.

Size (mm)	Time taken for cube to become yellow (s)			
	Trial 1	Trial 2	Trial 3	Trial 4
5 × 5 × 5	174	167	177	182
7 × 7 × 7	274	290	284	292
10 × 10 × 10	835	825	842	838

a) Name the process by which hydrochloric acid moves into the gelatine cubes in this experiment.

Diffusion

[1]

b) Calculate the average time taken for a 10 × 10 × 10 mm gelatine cube to become yellow.

835 s

[2]

c) Describe the relationship between the size of the gelatine cube and the time taken for the cube to become yellow.

The larger the cube, the longer it took for the cube to become yellow

[1]

[Total 4 marks]

2   Soaking an egg in vinegar dissolves its outer shell and leaves the egg surrounded by a partially permeable membrane. Some students use eggs which have been prepared in this way to investigate osmosis. They remove two eggs from the vinegar and place one of the eggs in a jar of water and the other in a jar of weak sugar solution.

a) Describe **one** way in which the students could measure the effect of osmosis.

[2]

b) Suggest what the results of this experiment would be.

[2]

c) Suggest a possible control for this experiment.

[1]

[Total 5 marks]

3   A student cut cylinders out of potatoes and placed them into different concentrations of sugar solution, as shown in the diagram.

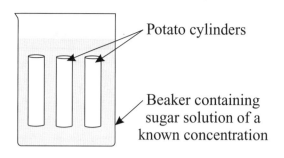

Potato cylinders

Beaker containing sugar solution of a known concentration

She measured the mass of the cylinders of potato before and after they had been placed in different concentrations of sugar solution for 20 minutes. Her results are shown below.

Concentration of sugar solution (M)	Change in mass of potato cylinder (g)			Mean change in mass (g)
	Cylinder 1	Cylinder 2	Cylinder 3	
0.0	+ 0.67	+ 0.65	+ 0.69	+ 0.67
0.2	+ 0.30	+ 0.31	+ 0.33	+ 0.31
0.4	+ 0.02	− 0.02	+ 0.01	0
0.6	− 0.27	− 0.31	− 0.25	− 0.28
0.8	− 0.48	− 0.5	− 0.47	− 0.48
1.0	− 0.71	− 0.65	− 0.72	− 0.69
1.2	− 0.78	− 0.81	− 0.82	− 0.80

a) Calculate the mean change in mass in a 1.2 M sugar solution.

− 0.80 g

[2]

b) Estimate the concentration of sugar inside the original potatoes.
Explain your answer.

0.4 M, because there was no change in mass when the potato was placed in this concentration of sugar, therefore the concentration of sugar inside the potato must have been equal.

[3]

c) Suggest why the student used three potato cylinders at each concentration of sugar solution and took a mean of the results.

To make the results more accurate

[1]

d) Suggest **one** variable that should have been controlled in this experiment.

The size of the potatoes

[1]

[Total 7 marks]

Exam Practice Tip

In your exams (particularly Paper 2), you could be given an unfamiliar experiment to answer questions about. Don't panic. Just apply your existing scientific knowledge (e.g. about how osmosis works and why you need to control a variable) and you'll be fine. Paying attention in your practical lessons will come in handy too.

Score

16

Active Transport

1   The diagram on the right shows a specialised plant cell.

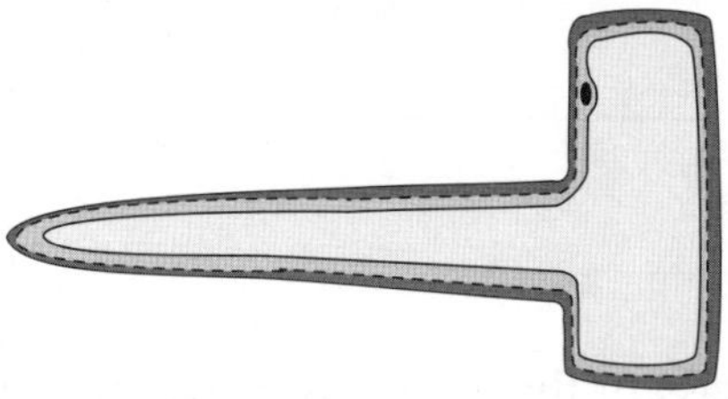

a) Give the name of this type of cell. ...*Root hair cell*...

                                            [1]

b) One of the main functions of this cell is to absorb mineral ions from the soil.

      Explain why minerals are usually absorbed from the soil by the process of active transport rather than diffusion.

      ..Because there is already a high concentration of sugars inside the root, so minerals need to be absorbed against the g: concentration gradient..

                                            [3]

c) i) This specialised cell contains many mitochondria. Suggest why.

          ...To provide energy needed for active transport............................

          ....

                                            [2]

      ii) Suggest **one** other way that this cell is adapted for absorbing lots of minerals from the soil.

          ...Has a large surface area to volume ratio.................................

                                            [1]

                                      [Total 7 marks]

2   Active transport can take place in the human body.

a) Explain why active transport is sometimes needed in the small intestine.

      ....

      ....

                                            [2]

b) Describe another example of when active transport is used in the human body.

      ....

      ....

                                            [2]

                                      [Total 4 marks]

Score:

11

Gas and Solute Exchange

1 Trout are freshwater fish. They are relatively large, multicellular organisms.
Euglena are small, single-celled organisms that live in both fresh water and salt water.

a) The rate of gas and solute exchange depends partly on the surface area to volume ratio
of the organism. Complete the following sentences.

Single-celled organisms have a relatively ......*large*............ surface area to volume ratio.

Larger multicellular organisms have a relatively ....*small*.......... surface area to volume ratio.

[1]

b) Explain why *Euglena* do not need specialised exchange organs, but trout do.

....

....

....

....

....

....

[4]

c) The gas exchange organs in fish are called gills.
Gas exchange takes place between the gill filaments and the water.

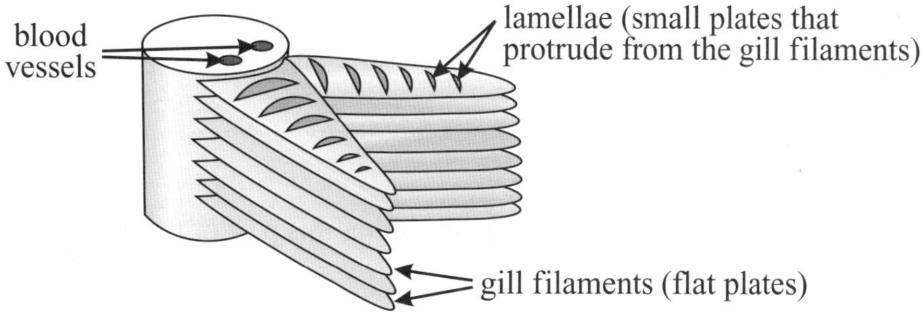

blood vessels — lamellae (small plates that protrude from the gill filaments) — gill filaments (flat plates)

Using your own knowledge and the diagram above, suggest and explain **two** ways in which a gill
is adapted to maximise the efficiency of gas exchange.

1. ....*Has a large surface area to volume ratio, so gas exchange can occur quickly*............

2. *Has lamellae*..

....

[4]

[Total 9 marks]

Score: ☐

9

Exchanging Materials in Mammals

1   Gas exchange in humans occurs in the alveoli. Alveoli are the tiny air sacs inside the lungs where gas exchange takes place.

a) State **three** features that make the alveoli well adapted for gas exchange.

1. ...Has a good blood supply...

2. ...Very thin walls...

3. ...large surface area to volume ratio..

[3]

b) Describe the gas exchange that occurs between alveoli and their surrounding blood vessels.

....

....

[2]

[Total 5 marks]

2   Coeliac disease is a disorder caused by the immune system reacting to a protein found in some foods, e.g. flour. The reaction can cause villi to become inflamed and flattened.

a) Where in the digestive system are villi found? *in the small intestine*........................

[1]

b) i) Give **two** structural features of normal villi.

1. ...Tall..

2. ...Thin..

[2]

ii) Suggest why it's an advantage to have villi lining this part of the digestive system rather than a flat layer of cells.

...To increase the rate of diffusion..

....

[1]

c) Suggest why coeliac disease sufferers might have low levels of some vitamins in their body.

...Because some vitamins would be too big to pass through...

...the inflamed villi...

....

[2]

Coeliac disease changes the shape of the villi — think about how this would affect their function.

[Total 6 marks]

Score: ☐

11

Genes and Chromosomes

1 Rank the following structures in order of size, from the smallest (**1**) to the largest (**4**).

> nucleus   gene   chromosome   cell

1 ..*gene*............   2 ...*chromosome*...   3 ...*nucleus*............   4 ....*cell*................

[Total 2 marks]

2 Complete the following passage using words from the box below.

> proteins   nucleus   genes   DNA   chromosomes   pairs

Most cells in your body contain a structure called the ...*nucleus*............ .

This structure contains chromosomes, which carry a large number of ...*genes*..................

and are usually found in ....*pairs*.................. .

[Total 3 marks]

3 The genetic material of a cell is made up of many different *genes*.

a) What is meant by the term *gene*?

..*a short section of DNA which controls the development of a specific characteristic*

[1]

b) Fur length in cats is controlled by a single gene.
A female cat gave birth to a litter of two kittens, shown on the right. One kitten is long-haired and the other is short-haired. Explain why the kittens show different characteristics.

> Think about how a single gene is able to produce different characteristics.

..*A gene is made up of alleles. The mother cat must have had one allele for long hair and one allele for short hair to enable her kittens to have different fur*

[3]

[Total 4 marks]

Score: ☐

9

😕 ☐   😐 ☐   🙂 ☐

Section One — Cells, Organs and Life Processes

Cell Division — Mitosis

1   Mitosis is a type of cell division.

a) Complete the following passage using words from the box.

once	different	four	destroys	copies	two	identical	twice

Mitosis is where a cell divides .....*once*................ to form ...*two*..................... new cells.

These new cells are ....*identical*........ to each other and to the parent cell.

Before a cell divides, it ...*copies*................ its DNA.

[4]

b) Give **two** reasons why cells divide by mitosis.

1. ...*For growth*...

2. *To repair old cells*...

[2]

[Total 6 marks]

2   The graph shows how the amount of DNA per cell changes as a cell undergoes two divisions by mitosis. Point **C** is the time when the chromosomes first become visible in the new cells.

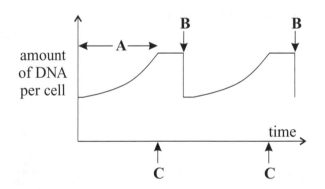

a) Describe and explain what is happening to the DNA during stage **A**.

.....*During stage A, the DNA is creating identical copies*..............

..*of itself*...

....

[2]

b) Suggest what happens at time **B**.

...*The cell divides*...

[1]

[Total 3 marks]

Score:

9

Cell Division — Meiosis

1 Mosquitoes have three pairs of chromosomes in their body cells. **Cell A,** shown to the right, is a mosquito cell which is about to divide by meiosis. **cell A**

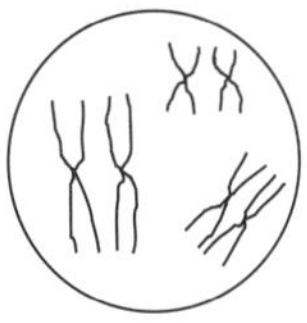

a) i) State the number of chromosomes in a cell
produced when cell A undergoes meiosis.

....

[1]

ii) How many divisions are there in the process
of meiosis?

...... ~~Rexen~~ Two ....

[1]

iii) How many cells will be produced in total
when cell A undergoes meiosis?

.... Four ....

[1]

b) The cells produced when cell A divides by meiosis are called gametes.
Suggest where cell A is likely to be located in the mosquito's body.

.... The reproductive organs ....

[1]

c) There is double the amount of DNA in cell A than in a normal mosquito body cell.
Explain why.

...... So that where the cell divides, the new cells have
the same number of DNA as all the other cell, *[1]*

[Total 5 marks]

2 At fertilisation, two gametes
combine to form a new individual.

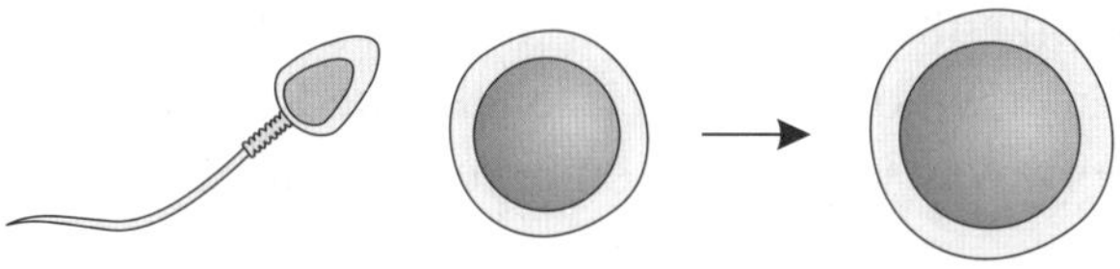

gametes combine   fertilised egg

Not to scale

a) Explain why it is necessary for gametes to have half the usual number of chromosomes.

..... So that when they fuse, the embryo has the correct
..... number of DNA e.g. the same amount from the mother
+ the father *[1]*

b) Describe how a single fertilised egg cell goes on to produce many different types of cells in the
new individual.

..... The egg cell divides by mitosis and the new cells
..... become specialised to perform a particular function ....

[2]

[Total 3 marks]

Exam Practice Tip

Ahh meiosis. And mitosis. Two very similar sounding words for two fairly similar processes — but it's important that you get the differences between them straight in your head for the exams. It might help you to remember that <u>me</u>iosis produces ga<u>me</u>tes (there's a 'me' in both of them). It's a good way of remembering the spelling too.

Score

8

Stem Cells and Differentiation

1 Complete the following passage about differentiation in different organisms.

Animal cells usually ...*lose*... the ability to differentiate at an early stage.

However, most ...*plant*... cells keep the ability to differentiate throughout their lives.

Undifferentiated cells that can develop into many different types of cell are called

...*stem cells*... .

[Total 3 marks]

2 Stem cells can be used to treat people suffering from some medical conditions. The stem cells can be used to replace faulty cells in the patient.

a) Give **one** example of a medical condition and state how stem cells could potentially be used to treat it.

...*Diabetes, stem cells can become insulin producing cells*...

[2]

b) Both adult and embryonic stem cells have the potential to be used for medical treatments. Read the following information about them.

> Adult stem cells are taken from a patient's own body, making them unlikely to be rejected when transplanted into the patient. They are relatively rare in adult body tissues however and can be hard to grow in culture. They can also only differentiate into a limited number of cell types.
>
> Embryonic stem cells have the potential to differentiate into any type of cell and are easy to grow in culture. There is a risk that embryonic stem cells could be rejected by the patient however, who would need to take medication to stop this from happening. Therapeutic cloning offers a way to produce embryonic stem cells that won't be rejected by the body.

In this question you will be assessed on the quality of your English, the organisation of your ideas and your use of appropriate specialist vocabulary.

Using the information above and your own knowledge, evaluate the potential use of adult and embryonic stem cells as medical treatments.

Hint: 'evaluate' basically means 'compare' here — so you need to make <u>clear comparisons</u> between the potential use of adult and embryonic stems cells in your answer.

....

....

....

....

....

....

....

[6] **Score:** ____

[Total 8 marks]

11

Growth of Tumours

1   Tumours can be benign or malignant. They are caused by cells dividing in an uncontrolled way.

a) State the purpose of normal, controlled cell division in mature animals.

To repair old cells

[1]

b) Describe the main difference between benign and malignant tumours.

benign tumours don't invade other cells, but malignant tumours invade other cells and form secondary tumours + are cancerous

[1]

c) Describe what happens in order for a secondary tumour to develop.

....

....

[2]

d) Tumours can be caused by chemical carcinogens and ionising radiation.

i) Name **two** sources of chemical carcinogens. *tobacco smoke, asbestos*

[2]

ii) Name **one** type of ionising radiation. *U.V. light*

[1]

[Total 7 marks]

2   A study was carried out to determine whether smoking is linked to an increased risk of breast cancer.

116 544 women without breast cancer in California were sent questionnaires to establish their smoking history and other personal information.
The women were then followed for 5 years.

The results on the right show the relative risk of breast cancer for women with different smoking histories.

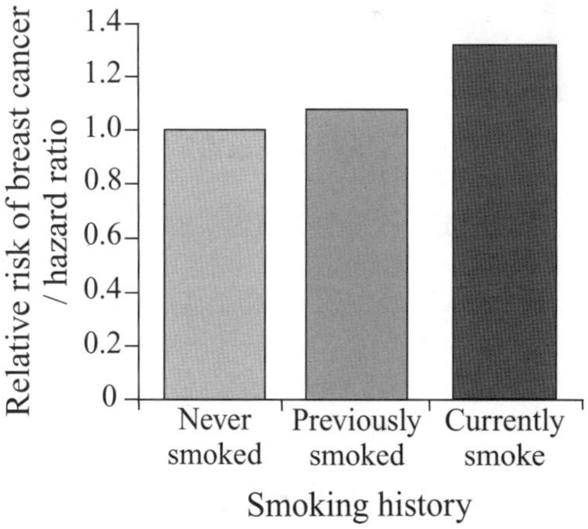

a) Describe the data shown in the graph.

women who have never smoked have a lower risk of breast cancer, whereas women currently smoking have a higher risk of breast cancer

[2]

b) What conclusion can you draw from these results?

If you smoke, you are more likely to get breast cancer (women)

[1]   **Score:**

[Total 3 marks]   **10**

The Respiratory System

1   The respiratory system is found in the thorax.

a)   Describe and explain the function of the respiratory system.

....

....

....

[3]

b)   A diagram of the human thorax is shown below.

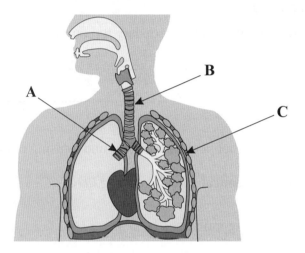

i)   Name the structures labelled **A-C** using words from the box.

intercostal muscle	diaphragm	bronchus	trachea	bronchiole

A ...*bronchus*...

B ...*Trachea*...

C ...*Intercostal muscle*...

[3]

ii)   Name the structure that separates the thorax from the abdomen.

   ...*Diaphragm*...

[1]

c)   Describe how muscles in the respiratory system cause a person to inhale.

...Intercostal muscles contract, pulling the ribcage...

up. The diaphragm contracts and flattens out. Thorax

volume increases which decreases the volume inside the

lungs causing air to be drawn in

Remember:
the diaphragm
is a muscle.

[4]

[Total 11 marks]

2 Emphysema is a disease of the lungs, commonly caused by smoking. In a person with emphysema, the walls of the alveoli are damaged. This reduces the alveoli's overall surface area. The capillaries surrounding the alveoli may also be destroyed.

Suggest and explain why a person with emphysema may have low oxygen levels in their blood.

....If....the....walls....of....the....alveoli....are....damaged,....this....means....
....that....the....is....no....longer....a....large....surface....area....to....volume....ratio +
....the....alveoli....can't....be....efficient....in....gas....exchange. Also....if....the....
....capillaries....are....damaged / destroyed....this....means....there....is....no....
....longer....a....good....blood....supply,....limiting....the....amount....of....oxygen
....in....the....blood + for....gas....exchange   *[Total 3 marks]*

3 An elderly woman is put on a positive pressure ventilator in hospital.

a) i) Explain how a positive pressure ventilator works.

....

....

[2]

ii) Suggest and explain **one** reason why the woman may have been put on an artificial ventilator.

....

....

[2]

b) Many years ago, if the woman had required the same treatment, she would have been put inside a negative pressure ventilator called an iron lung. An iron lung is a sealed case placed around a patient's chest.

i) When air is pumped out of the iron lung, the pressure around the patient's chest drops. Explain how this would cause air to be drawn into the patient's lungs.

....

....

[1]

ii) Suggest and explain how air is made to leave the patient's lungs using an iron lung.

....

....

[2]

[Total 7 marks]

Score:

21

Respiration

1 Respiration is carried out by all living cells. It can take place aerobically or anaerobically.

a) State the purpose of respiration.

....~~Getting oxygen to all cell~~ Release energy from glucose..........

[1]

b) Aerobic respiration takes place continuously in plants and animals.

   i) Fill in the spaces below to complete the balanced symbol equation for aerobic respiration.

   Tip: it might help you to write out the word equation somewhere first.

   $.C_6H_{12}O_6. + .6O_2. \rightarrow .6CO_2. + .6H_2O.$ (+ energy)

   [2]

   ii) Suggest why continuous aerobic respiration is beneficial to an organism.

   ....

   [1]

c) Describe how aerobic respiration helps plants to build proteins.

....

....

[2]

[Total 6 marks]

2 A rat is a mammal. It maintains a body temperature of around 37 °C, even in colder surroundings. A snake is a reptile. It is unable to maintain a constant body temperature.

Suggest and explain **one** reason why a rat may need to eat more food than a snake.

....

....

....

....

[Total 3 marks]

3 Diagram **A** below shows some muscle cells. Diagram **B** shows a skin cell.

Explain the difference between the relative numbers of mitochondria in each cell type.

Diagram A

Diagram B

mitochondria

mitochondria

....

....

....

....

Score:

[Total 3 marks]

12

Respiration and Exercise

1   Anaerobic respiration is less efficient than aerobic respiration.

a)   Suggest and explain when a person might start to respire anaerobically.

....

....

[2]

b)   Write the balanced symbol equation for anaerobic respiration in humans in the spaces below.

$$C_6H_{12}O_6 \rightarrow 2C_3H_6O_3 \quad (+ \text{ energy})$$

[2]

c)   Suggest **two** reasons why anaerobic respiration in humans is not efficient over long periods.

1.   *Build up of lactic acid* ....

2.   ....

[2]

d)   Plants can also respire anaerobically. Name the **two** substances produced during anaerobic respiration in plants.

.... *ethanol + carbon dioxide* ....

[2]

[Total 8 marks]

2   A scientist records the blood lactate level of a person at rest, immediately after 10 minutes of strenuous exercise, and at 10 minute intervals for the next 30 minutes. The results are shown in the table. The blood lactate level is a measure of the concentration of lactic acid in the blood.

a)   Explain why the blood lactate level is high immediately after exercise.

....

....

....

....

[2]

Time of record	Blood lactate level (mmol/Litre)
At rest, before exercise	1.5
Immediately after exercise	9.6
10 minutes after exercise	6.8
20 minutes after exercise	4.5
30 minutes after exercise	2.1

b)   Explain the changes in blood lactate level during the 30 minutes after exercise.

....

....

[2]

[Total 4 marks]

Section Two — Breathing and Circulation

3   A runner competes in a 1000 m race, which takes her 5 minutes.
    The graph below shows her breathing rate before, during and after the race.

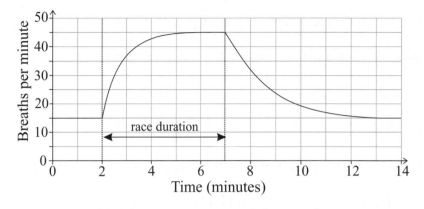

a) Calculate the increase in the number of breaths per minute during the race.

.... breaths per minute
[1]

b) Explain why exercise makes the runner's breathing rate increase.

    During muscle activity, more glucose + oxygen needs to be
    supplied to the muscle cells + extra CO_2 needs to be removed.
    Therefore, blood has to flow at a faster rate which means
    breathing rate increases
[3]

c) Explain why the runner's breathing rate does not return to normal immediately after the race.

    Because your lungs, heart + blood couldn't keep up with the
    demand for oxygen, you breathe heavier after exercise to get
    more O_2 into your blood + to remove lactic acid by oxidising it
    into CO_2 + water
[3]

d) Describe and explain **one** other way in which the runner's body will respond to exercise.

    Heart rate will increase, to supply more oxygen + glucose
    to the muscle cells quicker
[2]

[Total 9 marks]

4   The hormone glucagon is secreted by the pancreas. It stimulates the conversion of glycogen to
    glucose in the liver. Suggest why the blood glucagon level increases during vigorous exercise.

    During exercise glucose is being used up for energy. Therefore,
    blood glucagon level increases to convert more glycogen
    to glucose
[Total 2 marks]

Score: ☐

23

:(☐   :| ☐   :) ☐

Circulatory System — The Heart

1   The diagram shows the human heart and four blood
vessels, as seen from the front. The left ventricle
and pulmonary vein have been labelled.

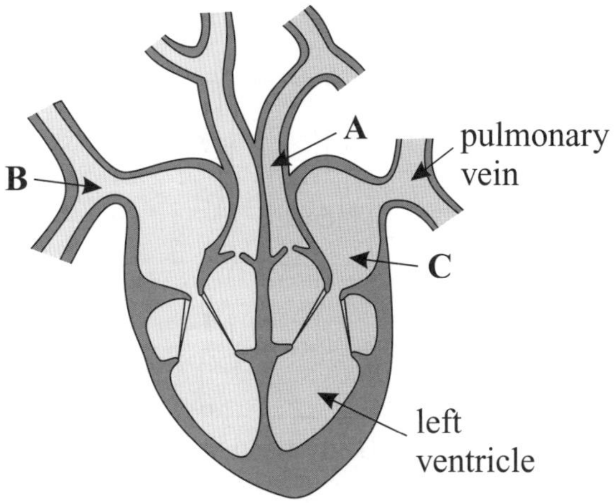

a) Name the parts labelled **A**, **B** and **C**.

A ...*Aorta*...

B ..*Vena Cava*.....................................

C ..*left atrium*....................................
[3]

b) i) Resting heart rate is controlled by a group of cells called the SAN. These cells act
as the heart's pacemaker. Where in the heart is this group of cells located?

...........*Right atrium*...
[1]

ii) If the SAN becomes defective, it can cause an irregular heart rate.
Suggest what could be done to correct this.

.......*use an artificial pacemaker*..
[1]

c) The tricuspid valve is a heart valve located between the right atrium and the right ventricle.
Suggest the function of the tricuspid valve.

.......*keeps blood flowing in the correct direction +*.................
.......*prevents backflow*...
[1]

d) Complete the sentence:

The walls of the ventricles are mostly made up of ......*muscle*.................. tissue.
[1]

e) Describe how deoxygenated blood from the body passes through the heart to reach the lungs.

....
....
....
....
....
[4]

[Total 11 marks]

2   The diagrams below show the circulatory systems of a fish and of a human.

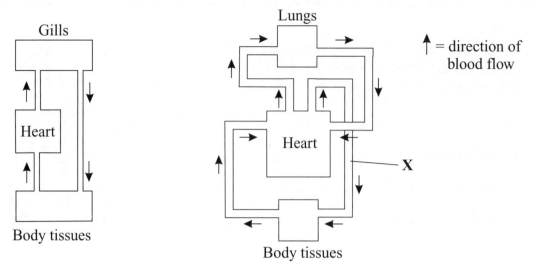

Fish Circulatory System   **Human Circulatory System**

a) Look at the diagram of the human circulatory system. Suggest whether blood vessels located at the point marked **X** are arteries or veins. Explain your answer.

   Arteries., carrying blood away from the heart

                                                                              [1]

b) Explain the purpose of a circulatory system.

   To carry oxygen around the body + to all the cells

                                                                              [2]

c) Use the diagrams and your own scientific knowledge to suggest:

   i) **two** differences between the circulatory system of a fish and that of a human.

      1. *A human has a double circulatory system whereas a fish doesn't*

      2. *In a fish, blood flows to the gills, but in a human blood flows to the lungs*

                                                                              [2]

   ii) **one** similarity between the circulatory system of a fish and that of a human.

      Blood flows to the body tissues

                                                                              [1]

                                                                    [Total 6 marks]

Exam Practice Tip

There's a lot to learn about the circulatory system and the heart. Knowing your way around a diagram of the heart is a good idea to help you understand how blood flows through it. Make sure you can identify each of the heart's chambers, as well as the main blood vessel associated with each one.

Score

17

Circulatory System — Blood Vessels

1 A diagram of a capillary is shown to the right.

a) Describe the function of a capillary.

...*Carries blood really close to*...
...*every cell in the body to*...
...*exchange substances with them*...

[2]

lumen

nucleus of cell

capillary

b) Suggest and explain **two** ways in which the structure of a capillary is adapted for its function.

1. ...*Walls that are one cell thick to provide a shorter diffusion path*...

2. ...*Has a large lumen to allow more blood to flow*...

[4]

[Total 6 marks]

2 A student did an experiment to compare the elasticity of arteries and veins. He dissected out an artery and a vein from a piece of fresh meat. He then took a 5 cm length of each vessel, hung different masses on it, and measured how much it stretched. His results are shown in the table.

a) i) The student was able to tell the difference between the arteries and veins in the meat because only veins contain valves. Explain the purpose of these valves.

...*Prevent back flow of blood / keep*...
...*blood flowing in the correct direction*...

[1]

ii) Suggest **one** other way in which the student would have been able to tell which blood vessels were arteries and which were veins when he was dissecting the meat.

...*Veins have a larger lumen than arteries*...

[1]

Mass added (g)	Length of blood vessel (mm)	
	Artery	Vein
0	50	50
5	51	53
10	53	56
15	55	59
20	56	-

b) Which vessel extended further? Suggest why this was.

...*Vein*, ~~has to have stretchy, elastic walls to allow a high~~ *has thinner walls, not as strong* ~~pressure of blood to flow through~~

[1]

c) The student plots the results of this experiment on a graph. Which variable should he put on the vertical axis? Explain your answer.

...*Length of blood vessel, because this depends on what mass is added*...

[2]

Score:

[Total 5 marks]

11

Circulatory System — The Blood

1 Plasma is one of the main components of the blood.

a) The plasma transports fragments of cells, called platelets, around the body.
State the function of platelets.

...Allow blood to clot...
[1]

b) Name **two** types of cell that are carried by the plasma.

..Red blood cells., white blood cell....................................
[2]

c) Name **two** other substances that are transported in the plasma.
For each substance, write down where it is transported from and to.

Substance 1: ..hormones...

Substance 2: ..antibodies...
[2]
[Total 5 marks]

2 The blood cell shown below transports oxygen around the body.

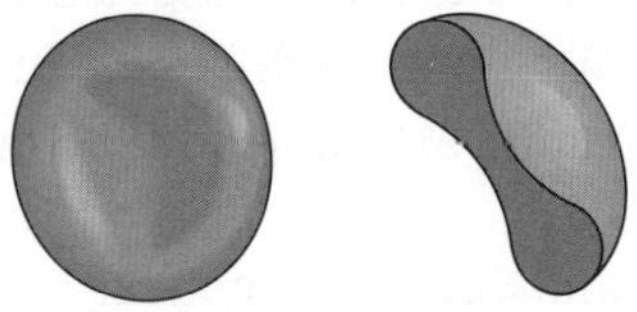

a) Give **one** way in which the internal structure of this cell is different to a blood cell involved in defending the body against disease.

..It has a concave structure.......................
....
[1]

View from above   Cut through view

b) Fill in the gaps to complete the sentences about the blood cell shown above.

This cell is packed full of a pigment called ....haemoglobin............. .

In the lungs, this pigment combines with oxygen to become ...oxyhaemoglobin.......... .

In the ......tissues............................, the opposite reaction happens and oxygen is released.
[3]
[Total 4 marks]

3 A person with Factor I deficiency may have little or no fibrinogen in their blood. Explain why a small injury such as a paper cut could be serious for a person with Factor I deficiency.

...Fibrinogen turns into fibre to form a mesh to allow............
.blood to clot. If there is little or no fibrinogen, blood can....
.not clot and blood will keep on flowing through the..........
.cut
[Total 3 marks]

Score:

12

Blood Type and Organ Donation

1 A donor kidney can be rejected by a patient's immune system.
This happens when the patient's antibodies attack the *antigens* on a donor kidney.

a) What is an *antigen*? ...*protein on the surface of every cell*...
[1]

b) Explain why a patient's antibodies may attack the antigens on a donor kidney.

...*Because the antigens are foreign / not recognise by the patients immune system*... [1]

c) Suggest **two** steps that should be taken to reduce the chances of rejection of a transplanted kidney.

1. ...*Tissue Typing*...

2. ...*Drugs should be taken to suppress the immune system*...
[2]

[Total 4 marks]

2 A person receiving a blood transfusion has to be matched with a suitable donor.

a) Complete the table to show which blood donors can give blood to which recipients.

Blood Group	Can give blood to:	Can get blood from:
A	A and AB	A and O
B	*B and AB*	*B and O*
AB	*only AB*	*anyone*
O	anyone	only O

To answer all of these question parts, you need to be thinking in terms of the antigens present on the red blood cells and the antibodies present in the blood.

[4]

b) Explain why a person with blood group A cannot receive blood from a person with blood group B.

...*Group A blood contains anti-B antigens, if it came in contact with group B blood it would cause agglutination*... [2]

c) Explain why a person with blood group O can donate their blood to a person with blood group A, B or even AB.

...*Group O blood doesn't have any antigens so cannot be attacked by group A, B or AB blood*... [2]

[Total 8 marks]

Exam Practice Tip

It can be tricky to get your head around all this antigens and antibodies stuff, let alone learn who can give blood to whom. But it's no good just learning the table off by heart and hoping for the best — take the time to really understand it all, then you'll be prepared for whatever the examiners throw at you in the exam.

Score

12

Circulation Aids

1 Complete the following sentences about coronary heart disease.

Coronary heart disease is when the ..... *arteries* ....

become blocked by ..... *fatty* .... deposits, narrowing them.

Stents are wire tubes that can be inserted into these blood vessels to keep them open,

making sure blood can pass through to the heart ... *muscle* ....

This keeps the cells supplied with .... *oxygen* ....

Sometimes you'll need to write more than one word to fill in the gap.

[Total 4 marks]

2 A faulty heart valve in a human could be replaced using a mechanical heart valve.

a) Describe **one** way in which damage to a heart valve may cause it to become faulty.

... *Blood could flow in both directions (leaky.)* ....
[1]

b) Describe **one** other way in which a faulty heart valve can be replaced.

..... *Biological heart valve* ....
[1]

[Total 2 marks]

3 A woman is suffering from heart failure. She is given a mechanical artificial heart to pump the blood around her body while she waits for a donor heart to become available.

In this question you will be assessed on the quality of your English, the organisation of your ideas and your use of appropriate specialist vocabulary.

Discuss the advantages and disadvantages of receiving an artificial heart compared to a donor organ.

....

....

....

....

....

....

....

....

[Total 6 marks]

Score:

12

Carbohydrates, Lipids and Proteins

1   Lipids are made of two types of smaller molecules joined together.

Name the **two** types of smaller molecules that lipids are made of.

1. .....*fatty......acids*..................................   2. ....*glycerol*..................................
[Total 2 marks]

2   Carbohydrates provide organisms with energy. They are made up of units of sugar.

a) Complete the sentences below.

Carbohydrates made of one or two units of sugar are called ....*simple*............... sugars.

.......*glucose*............. is an example of this type of carbohydrate.
[2]

b) Starch is a complex carbohydrate found in plants.

i) Give **one** other example of a complex carbohydrate.

...........*cellulose*...
[1]

ii) Describe the structure of a complex carbohydrate.

.....*a.....long....chain....of....simple....sugar....units....joined....together*...............
[1]
[Total 4 marks]

3   Proteins perform many different functions in the body.

a) Antibodies are proteins that help the body fight disease. They bind to molecules called antigens on the surface of foreign cells. The shape of an antibody must fit the shape of an antigen for it to bind. Explain how the structure of an antibody allows it to bind to a specific antigen.

..*Antibodies...are...made...of...long...chains...of...amino...acids.....These*.........

..*fold...into...a...specific...shape...to...allow...specific...antigens...to...fit*...

..*them*...
[2]

b) Give **three** other functions of proteins in the body.

1. .....*Structural.....proteins*..

2. ...*Hormones*...

3. ..*Enzymes*..
[3]
[Total 5 marks]

Score: [ ]

11

Enzymes

1   A student has a sample of an enzyme. She times how long it takes the enzyme to break down a substance at different pH levels. The results of the experiment are shown in the table below.

pH	time taken for reaction (seconds)
2	101
4	83
6	17
8	76
10	99
12	102

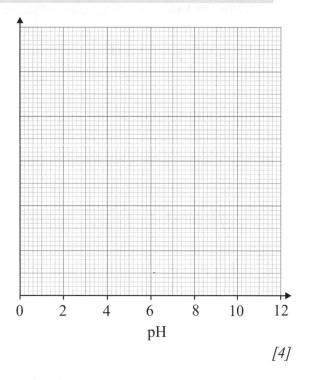

a) Draw a graph of the results on the grid provided.
Label the y axis and add an appropriate scale.
Draw a curve of best fit.

[4]

b) What is the optimum pH for this enzyme?

....

[1]

c) The pH of the reaction solution can break the bonds that determine an enzyme's shape.
Use this information to suggest why the reaction is very slow at certain pH levels.

....

....

[2]

d) Describe **two** things that the student should have done to make sure the experiment was a fair test.

1. ....

2. ....

[2]

e) Suggest how the speed of the reaction would differ if there was no enzyme present.
Explain your answer.

....

....

[2]

[Total 11 marks]

2 The graph shows the results from an investigation into the effect of temperature on the rate of an enzyme-catalysed reaction.

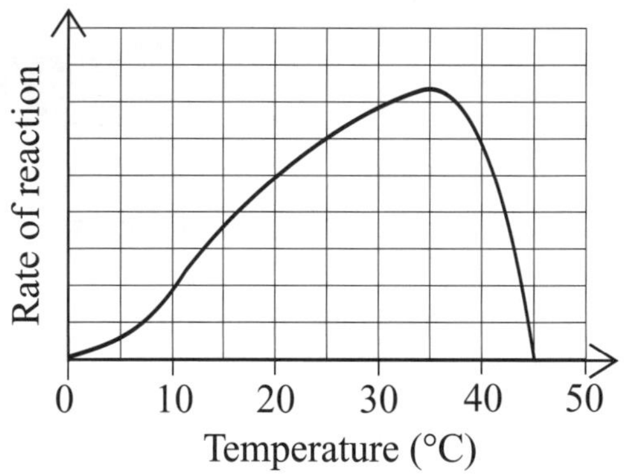

a) What is the optimum temperature for this enzyme?

....
[1]

b) Explain why the rate of the reaction is zero at **45 °C**.

....

....
[2]

[Total 3 marks]

3 Catalase is an enzyme that catalyses the breakdown of hydrogen peroxide, releasing oxygen. Catalase is present in many cells, including potato cells.

In this question you will be assessed on the quality of your English, the organisation of your ideas and your use of appropriate specialist vocabulary.

Describe an investigation to find out how temperature affects the rate of catalase activity in potato cells.

Your investigation should make use of the apparatus and materials shown in the diagram, as well as a stop watch and a water bath.

Your answer should include:

• what you would measure and how you would measure it
• how you would ensure the investigation was a fair test.

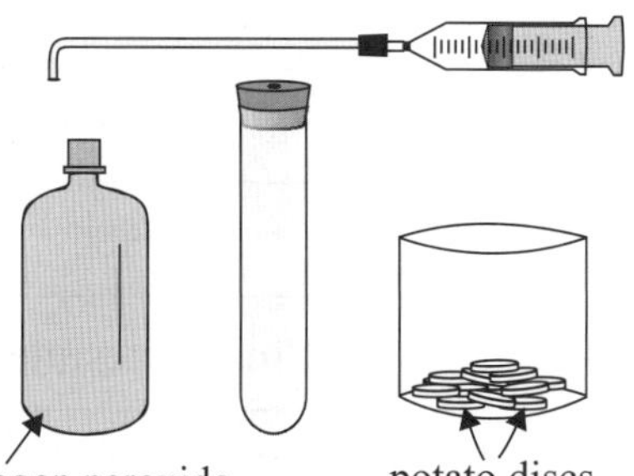

hydrogen peroxide solution   potato discs

The first thing you need to do here is to figure out how you could measure the rate of the catalase enzyme's activity. Then look carefully at the apparatus and materials you've got, and think how you could use them to set up an appropriate experiment.

....

....

....

....

....

....

....

....

....

.... **Score:** ☐

[Total 6 marks] **20**

Enzymes and Digestion

1 There are a number of digestive enzymes found in the human body.

a) Use the words from the box to complete the following sentences.

lipids	small intestine	amino acids	proteins	mouth

Protease converts ...*proteins*... into ...*amino acids*... .

Protease works in the ....*small intestine*... and the stomach.

[3]

b) State the role of amylase in digestion.

...*Breaks down starch into sugars*...

[1]

c) Different parts of the digestive system produce different enzymes. For each part of the digestive system, place a tick (✓) to show which enzyme or enzymes are **produced** there.

Location	Protease	Lipase	Amylase
Salivary glands			✓
Stomach	✓		
Small intestine	✓	✓	✓

Hint: some parts of the digestive system produce more than one enzyme.

[3]

[Total 7 marks]

2 Chronic pancreatitis is a condition where the pancreas becomes inflamed and cannot function properly.

a) A symptom of chronic pancreatitis is producing faeces with a high fat content. Suggest why this happens. Explain your answer.

....

....

....

[2]

b) Suggest **one** other effect chronic pancreatitis may have on digestion. Explain your answer.

....

....

[2]

[Total 4 marks]

3 Explain why molecules in the diet such as starch, proteins and lipids need to be broken down by digestive enzymes.

....

....

....

....

....

[Total 3 marks]

4 Gallstones are small, solid stones formed mainly of excess cholesterol. They can block the bile ducts (tubes) that connect the liver to the gall bladder and the gall bladder to the small intestine.

a) i) Name the digestive fluid stored in the gall bladder.

....*bile*..

[1]

ii) Name the organ in the body where this digestive fluid is produced, and where it acts on food.

Produced: ....*liver*..

Acts: ....*Small Intestine*..

[2]

iii) This digestive fluid is alkaline. Explain why this is important.

....*To neutralise stomach acid*..

....

[2]

b) Suggest and explain why eating fatty foods might cause a problem for people suffering from gallstones.

....

....

....

....

....

[4]

[Total 9 marks]

Exam Practice Tip

There are lots of simple facts you need to know about enzymes and digestion. Make sure you know what each enzyme or digestive fluid does, where it's produced and where it acts. In the exam you may need to apply this knowledge to an unknown scenario, so having these facts at your fingertips will make life much easier.

Score

23

More on Enzymes and Digestion

1 An exoenzyme is an enzyme that acts outside the cell where it's produced.
An endoenzyme is an enzyme that works within the cell where it's produced.

Are digestive enzymes exoenzymes or endoenzymes? Explain your answer.

....

....

[Total 2 marks]

2 The diagram shows the human digestive system.

a) Use the words from the box to name
the organs labelled **A**, **B**, **C** and **D**.

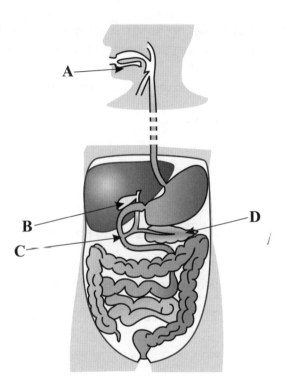

salivary glands	large intestine
anus	gall bladder
pancreas	stomach
oesophagus	duodenum

A ...*salivary glands*.............................

B ...*gall bladder*.............................

C ...*pancreas*..............................

D ...*duodenum*..............................

[4]

b) The conditions in the stomach are strongly acidic.

i) Name the substance the stomach produces to create this acidic environment.

.........*Hydrochloric acid*...

[1]

ii) Explain how acidic conditions aid digestion in the stomach.

.....*Hydrochloric acid provides the best pH for the protease*...........

...*enzyme in the stomach to work*..

[1]

c) Describe the role of the large intestine.

....*To absorb most of the water that is mixed with*.........

.*foods into the bloodstream*..

[1]   **Score:** [   ]

[Total 7 marks]   **9**

☹ [   ]   🙂 [   ]   😊 [   ]

Uses of Enzymes

1 Enzymes can be used in industrial processes used to produce foods.

a) Explain how and why enzymes are used in the manufacture of:

   i) baby foods.

   ....

   ....
   [2]

   ii) 'slimming' foods.

   ....

   ....

   ....
   [3]

b) Describe how carbohydrases are used in the food industry.

   ....
   [1]

   [Total 6 marks]

2 Biological washing powders contain enzymes which help to break down stains.

a) How are microorganisms involved in the production of biological washing powders?

   ....
   [1]

b) A manufacturer makes two different types of biological washing powder.
   Type **A** contains lipase enzymes. Type **B** contains protease enzymes.

   Which washing powder, **A** or **B**, would you recommend
   to someone who has got butter on their shirt? | A |

   Hint: think about what type of molecule butter is made from.

   Give a reason for your answer.

   ......It contains lipase enzymes which digest fats / lipids..................
   [1]

c) Using a biological washing powder can reduce the energy costs of doing laundry compared to
   using a non-biological washing powder. Suggest why this is.

   ......enzymes work best at lower temperatures / their optimum...
   ...temperature is lower than non-biological washing powders.
   A lower temperature saves energy ∴ is cheaper
   [2]

   [Total 4 marks]

   Score:
   10

The Nervous System

1 The nervous system allows humans to respond to *stimuli* and coordinate their behaviour.

a) What is meant by the term *stimuli*?

....*Stimuli are changes in the environment*..
[1]

b) Name the type of cells that detect stimuli in the sense organs. ...*receptors*.............................
[1]

c) Name the sense organ(s) that would contain cells to detect the following stimuli:

i) light ...*eyes*.............   ii) pressure ...*skin*.............   iii) changes in position ...*ears*.............
[3]

[Total 5 marks]

2 The information generated by a stimulus is transmitted to the central nervous system (CNS) as an electrical impulse. A coordination centre in the CNS then coordinates a response.

a) What type of cell does the impulse travel along to reach the central nervous system?

..........*Sensory neurone*..
[1]

b) Name the **two** parts of the body that make up the central nervous system.

1. ...*Brain*..

2. ...*Spine*..
[2]

c) Effectors respond to information from coordination centres.

Name **two** types of effector and describe how they respond to an impulse.

....

....

....

....
[4]

[Total 7 marks]

3 When a hungry dog smells food, its salivary glands start to produce saliva. Suggest what the stimulus, sense organ and effectors are in this scenario.

Stimulus: ....*Smell*...

Sense organ: ...*nose*...

Effectors: ....*salivary glands*...

Score:

[Total 3 marks]

15

Reflexes

1 A man picked up a plate in the kitchen without realising it was hot, then immediately dropped it. The diagram shows the reflex arc for this incident.

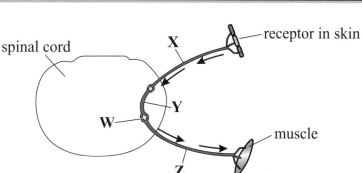

a) Name the **three** types of neurone labelled **X**, **Y** and **Z**.

X: ...*Sensory neurone*...

Y: ...*Relay neurone*...

Z: ...*Motor neurone*...

[3]

b) Name the coordination centre involved in this response.

...*Spinal cord*...

[1]

c) i) What name is given to the small gap between neurones, marked **W** on the diagram?

...*Synapse*...

[1]

ii) Suggest why the transmission of the nervous impulse is slowed down at this gap.

...*Because chemicals diffuse across the gap, which is a slow process*...

[2]

d) State what the effector is in this reflex arc and describe its response.

...*muscle, contracts to move away from the heat source*...

[2]

e) Explain why the man was able to drop the hot plate so quickly.

...*Because reflexes are unconscious activities, which are done automatically + the response is very quick*...

[2]

[Total 11 marks]

Exam Practice Tip

Reflexes always start with receptors (which detect the stimulus) and end with effectors (which carry out the response). They also tend to involve the same three types of neurone, in the same order. Get these basics learnt and you should be able to apply them to any reflex you happen to get asked about in the exams.

Score

11

The Brain

1 A hospital patient has a suspected brain injury. Doctors have been running tests on the patient's cerebral cortex, cerebellum and medulla.

a) The patient is breathing normally without any medical assistance. Using this information and the information above, suggest what part of the brain remains undamaged. Explain your answer.

.....The....medulla,...because...this...part...of...the....brain...controls.................

...unconscious....activities....e.g:..breathing...

[2]

b) Tests have shown that the patient's cerebral cortex has been damaged.

i) Give the letter, **X**, **Y** or **Z** that corresponds to this region of the brain on the diagram.

X

[1]

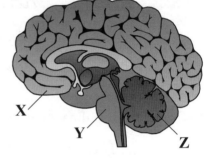

ii) State **two** activities that are associated with this region of the brain.

1. ...Intelligence......,.....memory...

2. ..language......,.....consciousness...

[2]

[Total 5 marks]

2 OPCA is a condition that affects the function of particular areas of the brain. Symptoms include clumsiness, difficulty walking and loss of balance.

a) Using the information given above, suggest a region of the brain that may not be functioning normally in people suffering from OPCA. Explain your answer.

..The......cerebellum,.....because...this...part...of...the....brain.......................

...controls.....muscle.....coordination...

[2]

b) State **three** methods scientists can use to work out the function of particular parts of the brain.

1. ..Electrically...stimulating...the...brain...

2. ...MRI...scans...

3. .Studying...patients.....with...brain...damage...

[3]

[Total 5 marks]

Score:

10

😐 ☐   🙂 ☐   😉 ☐

Homeostasis

1 A cyclist goes for a two hour bike ride. It is a hot day and he has to work hard on some steep hills. Automatic control systems help to regulate the water content in his body.

a) The cyclist loses approximately 1.2 L of water per hour through his skin in sweat.

i) His water bottle holds 0.8 L of water. Calculate how many full bottles of water the cyclist will have to drink in order to replace the water he loses in his sweat during the bike ride.

$$1·2 \times 2 = 2·4$$
$$2·4 \div 0·8 = 3$$

..............3.............. bottles
[2]

ii) Give **one** other way in which the cyclist's body will lose water while he is cycling.

......Breathing....via....the....lungs..
[1]

b) The water content of the cyclist's body is controlled by hormones.
How are hormones transported to their target organs? ......In....the....blood........
[1]

c) Instead of water, the cyclist fills his water bottle with a sports drink. The sports drink contains water, sugar and ions. Suggest why drinking a sports drink may be more beneficial for the cyclist than drinking water alone.

....The....sports....drink....will....provide....the....energy....needed....for............
....exercise.....The....cyclist....will....lose....ions....in....his....sweat....+....glucose........
....which....can....be....replaced....by....the....sports....drink   [2]

[Total 6 marks]

2 A woman is on holiday in Norway. It is –5 °C and she is feeling cold. A coordination centre in her central nervous system receives and processes information about her body temperature from receptors. The coordination centre then sends electrical impulses to effectors in her body.

a) Name the coordination centre that is responsible for monitoring body temperature.

....The....Thermoregulatory....centre..
[1]

b) Give the **two** locations of receptors in the automatic control system described above.

1. ....Her....skin..
2. ....Thermoregulatory....centre....in....the....brain............................
[2]

c) Explain why this automatic control system is activated when the woman feels cold.

....To....try....and....keep....her....body....temperature....constant
[1]   **Score:**

[Total 4 marks]   **10**

The Liver and Kidneys

1   Some important chemical reactions take place in the liver.

a)   The liver breaks down excess amino acids, which can be damaging if they remain in the body. Describe the process by which amino acids are broken down ready for excretion.

...........Deamination...

..

..

[3]

b) Poisonous substances, such as alcohol, are broken down into less harmful substances in the liver.

i)   What name is given to this process?

..........detoxification...

[1]

ii) How are the products of this process removed from the body?

.......In.....the.....urine.....via.....the...kidneys..

[1]

c)   Old blood cells are also broken down in the liver. What product of this process is stored in the body for later use?

...............Iron...

[1]

[Total 6 marks]

2   In the kidneys, blood is filtered and waste substances are removed for release in the urine.

a)   Name **two** substances that are found in a healthy person's urine.

1. *...Water...*

2. *...amino...acids...*

[2]

b)   A doctor carries out tests to see if a patient's kidneys are functioning properly. She finds that the patient has a normal blood glucose level, but a high level of glucose in his urine.

Explain how this could indicate that the patient's kidneys are not functioning properly.

..

..

[2]

c)   Sodium is a dissolved ion that is filtered out of the blood in the kidneys. In a healthy kidney, what determines the amount of sodium ions that are reabsorbed back into the blood?

..

[1]   **Score:**

[Total 5 marks]   **11**

Controlling Water Content

1   Drinking alcohol can reduce the production of the hormone ADH. This can lead to dehydration (a condition in which the body does not contain enough fluid).

  a) Name the part of the brain where ADH is produced.

......*The pituitary gland*..

[1]

  b) Explain how reduced ADH production could lead to dehydration.

....

....

....

[3]

[Total 4 marks]

2   A runner went for a 10 mile run on a warm day.

  a) When she got home she noticed that her urine was darker in colour than normal. Explain why she produced darker coloured urine.

....

....

....

....

....

[4]

  b) The next day the runner did another 10 mile run on the same route. When she got home she noticed that her urine was paler than it had been after her run on the previous day. Suggest **one** possible reason for this and explain your answer.

....

....

....

....

....

[4]

[Total 8 marks]

Score:

12

Kidney Failure

1 Kidney failure can be treated by dialysis or a kidney transplant.

Tick (✓) the boxes in the table to show which features apply to each type of treatment.

Feature of treatment	Dialysis	Transplant
High risk of infection		
Long-term, one-off treatment		
Patient needs to visit the hospital several times per week		
Patient can lead a relatively normal life		

[Total 5 marks]

2 A model of dialysis is shown below. No movement of substances has taken place yet.

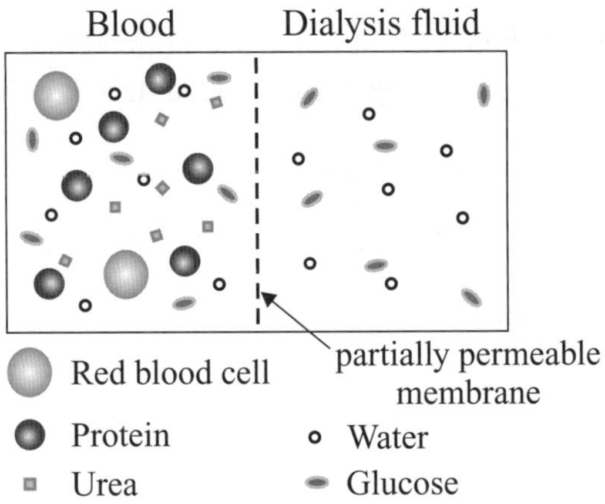

a) Proteins and red blood cells will **not** diffuse across the membrane from the bloodstream into the dialysis fluid. Explain why.

....

....

[1]

Blood   Dialysis fluid

Red blood cell   partially permeable membrane

Protein   ○ Water

▪ Urea   ● Glucose

b) i) Describe the difference between the concentrations of glucose on either side of the membrane. Suggest a reason for this.

....

....

[2]

ii) Describe the difference between the concentrations of urea on either side of the membrane. Suggest a reason for this.

....

....

[2]

c) Explain why dialysis needs to be carried out regularly.

....

....

[1]

[Total 6 marks]

3   Some information is given below about the number of patients
on the waiting list for kidney transplants in the UK.

	On 31/03/12	**On 31/03/13**
Total number of patients waiting for a kidney transplant	6633	6348

Number of patients receiving a kidney transplant between 01/04/12 and 31/03/13: **2998**

a) i) What conclusions can you draw from this information?

....

....

[2]

ii) Suggest **one** reason for the difference between the number of patients waiting for a kidney
transplant on 31/03/12 and the number of patients receiving a transplant over the following
12 months.

....

[1]

Patients with kidney failure can be treated using dialysis instead of a kidney transplant.
The table below shows some of the costs associated with each treatment.

Treatment	**Average cost per patient (£)**
Dialysis	30 800 per year
Transplant	17 000
Anti-rejection drugs	5000 per year

UK adults with kidney failure can wait several years for a kidney transplant.

b) Calculate the amount of money saved per patient when a transplant is performed instead of
three years of dialysis.

> Don't forget that transplant patients
> need drugs to stop organ rejection.

Money saved ....

[2]

c) Suggest **one** other benefit of having a kidney transplant rather than regular dialysis.

....

....

[1]

[Total 6 marks]

Score:

17

Controlling Body Temperature

1 Changes in the skin are an important part of temperature regulation.
   The diagram below shows a cross-section through the skin of a person who is cold.

Use the diagram and your own knowledge to describe
how the blood vessels close to surface of the skin
respond to the cold. Explain this response.

....

....

....

[Total 3 marks]

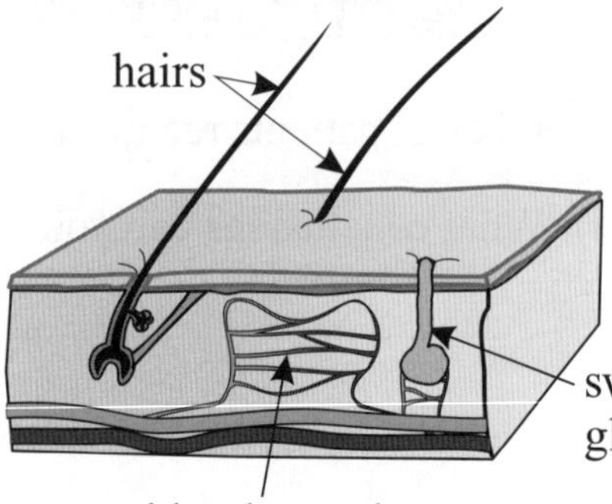

hairs

sweat
gland

blood vessels

2 Temperatures in a sauna can reach up to 100 °C.

a) Explain why a person using a sauna may be advised to drink water regularly.

....

....

[2]

b) Suggest how the blood vessels close to the surface of the skin would respond to being in a sauna.
    Explain your answer.

....

....

....

[3]

[Total 5 marks]

3 Like many other mammals, seals shiver in response to the cold. However, a study
   of hooded seals found that when they dive in cold water, shivering is inhibited.

a) Explain how shivering can help to keep seals warm.

....

....

....

[3]

b) Suggest how **not** shivering could benefit seals that are diving for long periods.

....

....

....

Hint: like all mammals,
seals can't breathe
underwater. So they
need to conserve oxygen
when they dive.

[3]   **Score:**

[Total 6 marks]   **14**

Controlling Blood Glucose

1 The concentration of glucose in the blood is constantly monitored by the body. Automatic control systems are in place to make sure the concentration doesn't become too high or too low.

a) Name the organ that monitors the blood glucose concentration. ....

[1]

b) Complete the following sentences.

If the level of glucose in the blood becomes too high, ... is released

by the .... This causes glucose to move from the blood into the ....

and the blood glucose level is ....

[4]

c) Explain how the blood glucose level is controlled when there is not enough glucose in the blood.

....

....

....

....

[3]

d) Excess glucose is sometimes stored by the body as lipid.
In what circumstances does this happen?

....

[1]

[Total 9 marks]

2 Glycogen synthase is an enzyme, which helps to convert glucose into glycogen in the body.

a) Glycogen synthase is present in the liver.
Give **one** other location in the body where glycogen synthase is also likely to be present.

....

[1]

b) Some people do not have enough functioning glycogen synthase in their liver.
When these people eat, their blood sugar level can become extremely high. Suggest why.

....

....

....

[3]

c) Suggest the name of a hormone that increases glycogen synthase activity.

....

[1]

[Total 5 marks]

3   A man has Type 1 diabetes. He treats his condition by regularly injecting himself with insulin.

a) Explain why it is necessary for the man to regularly inject insulin.

....

....
                                                                                                                          [2]

b) The man was asked by his doctor to monitor his blood glucose level over the course of a day.
    The graph below shows some of the data he collected.

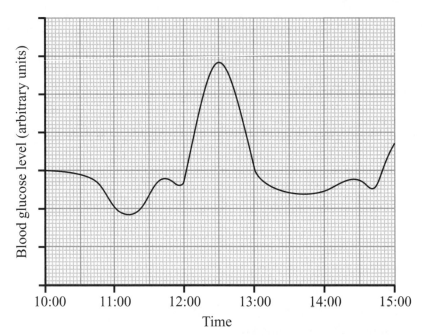

In the time period shown on the graph the man injected himself with insulin once. Using the
information in the graph, suggest what time the man injected insulin. Explain your answer.

....

....
                                                                                                                          [2]

c) i) A woman has Type 2 diabetes. She is unable to control her condition by injecting insulin.
        Explain why, and suggest how different medication could help her to control her diabetes.

    ....

    ....

    ....
                                                                                                                          [3]

   ii) State a health problem that can significantly increase the chance of developing Type 2 diabetes.

    ....
                                                                                                                          [1]

d) Apart from the methods described above, give **one** way in which diabetes can be controlled.

....
                                                                         [1]   **Score:**

                                                                     [Total 9 marks]   **23**

Fighting Disease

1 Bacteria and viruses can cause disease and make you feel ill.

a) Give **one** way that bacteria can make you feel ill.

...... They produce toxins ....
[1]

b) Explain how viruses can make you feel ill.

...... They invade healthy cells, reproduce + the cell bursts.
.... releasing the viruses, making you feel ill ....
[2]

[Total 3 marks]

2 The body can defend itself against pathogens.

a) What is a *pathogen*?

...... Foreign microorganisms ....
[1]

b) i) Use words from the box to complete the sentences.

pathogens	all	antibodies	ingesting	toxins	particular	fighting

White blood cells can produce ... antibodies ....

These can destroy ..... pathogens ....

One type of antibody will destroy ..... particular .... bacteria or viruses.

White blood cells can also destroy bacteria or viruses by .... ingesting .... them.
[4]

ii) Describe how antitoxins help defend the body against pathogens.

...... Antitoxins counteract the toxins produced by ....
.... pathogens ....
[1]

c) The chickenpox virus is a pathogen. People who have had chickenpox once do not usually suffer any symptoms of the disease if they are exposed to the pathogen again. Explain why.

....
....
....
[2]

[Total 8 marks]

Score: [ ]

11

Fighting Disease — Vaccination

1 The MMR vaccine is available to all children in the UK. The vaccine is given to a child as a baby and then again around the time the child starts school.

a) i) Complete the following sentences.

The MMR vaccine protects against measles and two other diseases. Once someone has had the

vaccine they will be .... to these diseases.

[1]

ii) Give the names of the **two** diseases, other than measles, that the MMR protects against.

1. ....

2. ....

[1]

b) Two children are exposed to the measles virus.

Child A has had the MMR vaccine and does not develop measles.
Child B has not had the MMR vaccine and does develop measles.

Explain why.

....

....

....

....

....

....

[4]

c) The MMR vaccine has a small risk of serious side effects, including seizures and allergic reactions. However, the government recommends that all children are given the MMR vaccine.

Suggest **two** reasons why the government still recommends that all children are given the MMR vaccine.

1. ....

....

2. ....

....

[2]

[Total 8 marks]

Score:

8

Fighting Disease — Drugs

1 A doctor told a patient with a cold to take some painkillers.

a) Explain why the painkillers will not cure the cold, but may help the patient to feel better.

....

....
[2]

b) The patient asked the doctor for some antibiotics.

i) Name **one** antibiotic. ....
[1]

ii) The doctor said that antibiotics would not help the patient. Suggest why not.

....

....
[2]

[Total 5 marks]

2 The graph shows the number of bacteria in a patient's blood during a course of antibiotics.

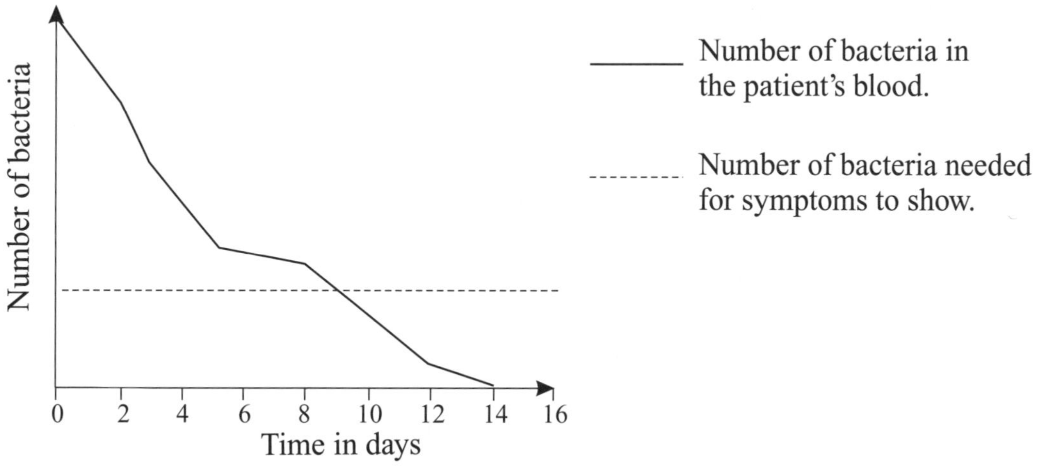

——— Number of bacteria in the patient's blood.

- - - - - Number of bacteria needed for symptoms to show.

a) How long after starting the course of antibiotics did the patient's symptoms disappear?

....
[1]

b) Explain why it is important that the patient was given the right type of antibiotic for his infection.

....
[1]

c) There is a wide range of antibiotics that are used to treat bacterial infections.
Explain why it is harder to develop drugs to treat viruses.

....

....
[2]

[Total 4 marks]

52

3 Antibiotics have been widely used since the 1940s.

a) Describe what has happened to the number of deaths from infectious bacterial diseases since the introduction of antibiotics.

....

[1]

b) In recent years, antibiotic-resistant strains of bacteria have become more common.

i) Give **two** ways that humans have increased the rate of development of resistant strains.

1. ....

2. ....

[2]

ii) Give **two** ways in which people are tackling the problem of resistant strains.

1. ....

2. ....

[2]

[Total 5 marks]

4 *Staphylococcus aureus* (SA) is a common bacterium that can enter the body through open wounds. It may cause mild infections in healthy people, but more serious illness in people with weakened immune systems. Some strains of SA have developed resistance to the antibiotic methicillin, and are known as methicillin-resistant *Staphylococcus aureus* (MRSA).

a) Suggest why MRSA is a more serious problem in hospitals than in wider society.

....

....

[2]

b) i) Explain how a population of methicillin-resistant *Staphylococcus aureus* could have developed.

....

....

....

....

[4]

ii) Suggest and explain how a new strain of antibiotic-resistant bacteria could lead to a big outbreak of disease.

....

....

....

[3]

[Total 9 marks]

Score: ☐

23

Section Five — Defending Against Disease

Investigating Antibiotic Action

1 A student is asked to carry out an experiment to investigate which disinfectant, from a choice of five, is most effective against bacteria found on the work surface of the laboratory.

a) *In this question you will be assessed on the quality of your English, the organisation of your ideas and your use of appropriate specialist vocabulary.*

Describe an experiment the student could do to investigate which disinfectant is most effective against the bacteria on the work surfaces of the laboratory.

Your answer should include:

- the method and equipment the student would use.

- how the student would make the investigation a fair test.

- any safety precautions the student should take.

Hint: as part of your answer to this question, you should write about how you would avoid contaminating your experiment with unwanted microorganisms.

....

....

....

....

....

....

....

....

....

....

....

....

[6]

b) Suggest **one** reason why the results of this experiment might be produced more quickly in industrial conditions than in a school laboratory.

....

....

[2]

[Total 8 marks]

Exam Practice Tip

If you get asked to describe an experiment in the exam, think about how to organise your answer before you start. If you're told that you'll be assessed on the quality of your English, then you'll be marked on how well you write — so make sure your writing is legible and be careful with your spelling, punctuation and grammar.

Score

8

Section Six — Plants

Photosynthesis

1 In plants, photosynthesis is used to produce sugar (glucose).
It can be represented by word and symbol equations.

$$\text{carbon dioxide} + \mathbf{A} \xrightarrow{\text{light energy}} \text{glucose} + \mathbf{B}$$

a) i) Name the reactant in photosynthesis represented by **A** in the equation above.

...... Water ....

[1]

ii) Name the by-product of photosynthesis represented by **B**.

.... Oxygen ....

[1]

b) Complete the balanced symbol equation for photosynthesis.

$$.6CO_2.... + ..6H_2O.... \xrightarrow{\text{light energy}} ..C_6H_{12}O_6.... + ..6O_2....$$

[2]

c) Where does the carbon dioxide involved in photosynthesis come from?

.... The air ....

[1]

[Total 5 marks]

2 A substance called chlorophyll is needed for photosynthesis to take place.
Chlorophyll is present inside algae so they can carry out photosynthesis.

The diagram on the right shows an algal cell.

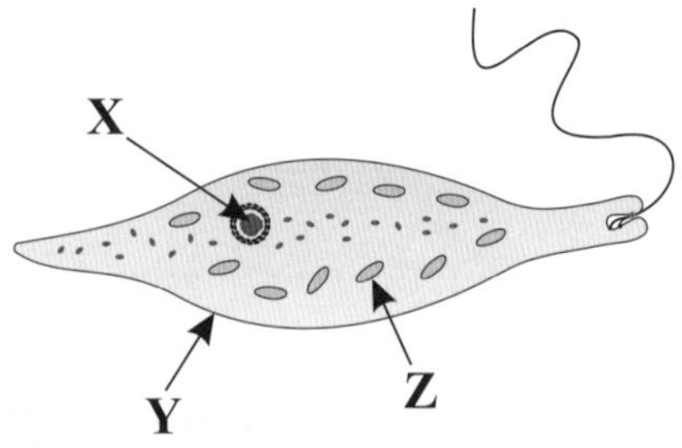

a) i) On the diagram, which letter, **X**, **Y** or **Z**,
labels a structure where chlorophyll is found? [Z]

[1]

Remember, an algal cell is similar in structure to a plant cell.

ii) Name the structure where chlorophyll is found.

.... Chloroplasts ....

[1]

b) Explain why chlorophyll is needed for photosynthesis.

..... Chlorophyll absorbs light energy for photosynthesis ....

....

[1]

[Total 3 marks]

Score: [   ]

8

The Rate of Photosynthesis

1 A farmer in the UK doesn't put his cows out during the winter because the grass is not growing.

Suggest **two** reasons why the growth rate of grass in the UK is lower in winter than in summer. Explain your answer.

There is less sunlight which means that the rate of photosynthesis is slower. Also, there are less hours of daylight in the winter as it gets darker earlier, which means that grass can't photosynthesise as often as it can in summer

[Total 4 marks]

2 Seth investigated the effect of different concentrations of carbon dioxide on the rate of photosynthesis of his Swiss cheese plant.

Graph axes: *rate of photosynthesis* (y-axis) vs *light intensity* (x-axis)

Legend:
- $0.1\% \; CO_2$
- $0.07\% \; CO_2$
- $0.04\% \; CO_2$

The results are shown on the graph.

a) Describe the effect that increasing the concentration of CO_2 has on the rate of photosynthesis as light intensity increases.

Light intensity is a measure of the brightness of the light that the plant is exposed to.

At low light intensity, increasing CO_2 has no effect, but at higher light intensity, increasing CO_2 increases the maximum rate of photosynthesis

[2]

b) Explain why all the lines on the graph level off eventually.

Because there is another limiting factor e.g. CO_2 or temperature

[1]

[Total 3 marks]

3 Sunlight contains light of different wavelengths, some of which we see as different colours. The amount of light absorbed at each wavelength for the pigment chlorophyll is shown below.

a) What colours and wavelengths of light are best absorbed by chlorophyll?

Blue, wavelength = 450 mm
Red, 660 mm

[2]

b) Suggest how you could use the information on the graph to increase the growth rate of plants in a greenhouse.

Use an artificial blue light

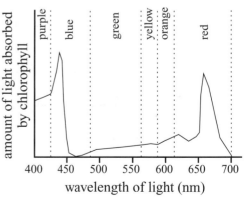

[1]

[Total 3 marks]

4 Greenhouses make it easier for farmers to control the environment around plants. Some farmers use paraffin heaters inside their greenhouses. The word equation for the burning of paraffin is: paraffin + oxygen → carbon dioxide + water (+ heat).

Explain the benefit to farmers of using paraffin heaters inside their greenhouses.

Paraffin heaters give a constant supply of carbon dioxide and increase the temperature inside a greenhouse. Therefore the rate of photosynthesis will be higher meaning that the plants will grow faster and bigger

[Total 4 marks]

5 A student investigated the volume of oxygen produced by pondweed at different intensities of light. Her results are shown in the table.

Relative light intensity	2	4	6	8	10	12
Volume of oxygen evolved in 10 minutes (cm³)	12	25	15	48	61	72

a) What was the student trying to work out by measuring the volume of oxygen produced?

Rate of photosynthesis

[1]

b) Plot a graph of the results on the grid provided.

[3]

Relative light intensity

c) One of the results is anomalous. Give the light intensity at which this result occurred and suggest an error that the student could have made to cause this result.

6, the light intensity could have been incorrect

[2]

d) The results show that the rate of photosynthesis increases as light intensity increases.

Describe and explain what would have happened to this relationship if the light intensity in this experiment had been increased further.

Eventually, the graph would level off because light is no longer the limiting factor

[2]

[Total 8 marks]

Score: ☐

22

How Plants Use Glucose

1 The glucose produced by photosynthesis has a variety of uses within plants.

a) Complete the following passage using words from the box.

lipids	respiration	transpiration	cellulose

Some of the glucose produced by photosynthesis is used for ......*respiration*.........,

which releases energy for the plant. Glucose is also used to produce ......*cellulose*............,

which can be used to strengthen cell walls.

[2]

b) Some of the glucose produced by photosynthesis is converted into starch for storage.

i) Name another storage molecule used in plants, which can be produced using glucose.

...............*fats / oils / lipids*.............................

[1]

ii) Starch is made up of many individual glucose molecules linked together.
It can be broken down to provide a source of glucose when needed.
Suggest when a plant may need to access stored glucose in this way.

........*During winter, when there is less light for photosynthesis*

[1]

c) Explain why starch is used as a storage molecule rather than glucose.

.........*Starch is insoluble, doesn't cause water to be drawn*
.....*into the cells via osmosis*

[2]

[Total 6 marks]

2 Sundews are carnivorous plants, which trap insects with a sticky secretion.

a) Sundews use glucose and nitrate ions to make proteins.

i) How do sundews obtain glucose?

.....*Through photosynthesis*

[1]

ii) Sundews have a method of obtaining nitrate ions that most plants cannot use.
Explain how and why this method is different to the one used by most plants.

....

....

....

[3]

b) Give **one** other example of a carnivorous plant. ......*Venus Fly Trap*......

[1]

Score: ☐

[Total 5 marks]

11

Exchanging Materials in Plants

1 Leaves and roots are important exchange organs in plants.

a) i) Give **two** substances that roots can absorb from the soil.

   ....

   [2]

   ii) Explain how roots are adapted for absorbing substances from the soil.

   ....

   ....

   [2]

b) Give **two** ways in which the surface area of a leaf is increased for effective gas exchange.

   ....

   [2]

   [Total 6 marks]

2 Carbon dioxide from the atmosphere enters plants through their leaves.

a) Name the process by which carbon dioxide enters the leaves.

   ....

   [1]

b) Carbon dioxide enters leaves through the stomata.

   i) On the diagram, label **one** of
      these structures with the letter **X**.

   [1]

   ii) Name the by-product of photosynthesis that exits leaves through the stomata.

   ....

   [1]

c) A group of scientists have discovered a plant mutant with a defect in its guard cells.
   The defect means that the guard cells do not function normally.

   Five mutant plants and five non-mutant plants are placed into a hot, dry, well-lit environment for
   3 hours. During this time, the soil the plants are grown in is kept well-watered. After 3 hours the
   mutant plants have wilted, whereas the non-mutant plants appear normal.

   Suggest how the guard cells in the mutant plants are defective. Explain your answer.

   ....

   ....

   ....

   ....

   [3]

   [Total 6 marks]

3   A scientist measured the rate of water loss in two plants over 48 hours.
     The results are shown in the graph.

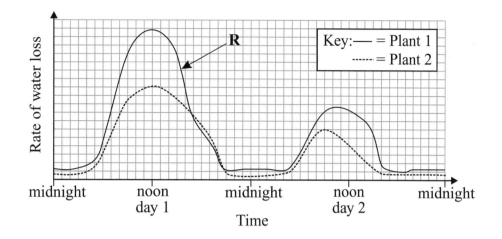

a) At what time on **day 2** was the rate of water loss highest for **plant 2**?

     ....
                                                                                                        [1]

b) The rate of water loss for both plants was slower on day 2 than on day 1.
     Suggest **two** reasons for this.

     1. ....

     2. ....
                                                                                                        [2]

c) At point **R** on the graph, **plant 1** was wilting.
     Suggest an explanation for this.

     ....

     ....
                                                                                                        [1]

d) In a second experiment, the scientist covered the   Petroleum jelly is a
     undersides of the leaves with petroleum jelly.   waterproof substance.
     Explain how this would affect the rate of water vapour loss from the leaves.

     ....

     ....

     ....
                                                                                                        [3]
                                                                                            [Total 7 marks]

Exam Practice Tip   Score

Examiners love a good graph, so make sure you're happy with interpreting one. If you're asked to read a
value off a graph, take a good look at the axes first and make sure you understand exactly what's being
shown. It's always a good idea to include the correct units in your answer too. You could pick up a mark.

                                                                                                        19

Water Flow Through Plants

1   Diagram **A** shows a cut flower stem in a beaker of water-soluble dye.
Diagram **B** shows the same flower after a period of several hours.

Tip: flower petals have stomata, just like leaves.

    a)   Describe and explain what has happened to the
dye to produce the results shown in diagram **B**.

....

....

....

....

....   *[3]*

A   **B**

    b)   i)   Suggest how you could adapt the experiment shown above to measure the effect of temperature
on the rate of transpiration in the cut flower stem.

....

....   *[1]*

      ii) Suggest **two** variables you would need to control to make an experiment into the effect of
temperature on transpiration rate a fair test.

1. ....

2. ....   *[2]*

[Total 6 marks]

2   When a tree is 'girdled', a ring of bark is removed from around the tree trunk (along with the
outer layer of wood), as shown in the diagram. The phloem are located around the outside of
a tree trunk only, so girdling completely removes the phloem in the girdled area. Girdling is
sometimes used to remove a single tree from an area without damaging other vegetation.

girdle
bark
removed
tree trunk

Using the information above and your own scientific knowledge,
suggest and explain what effect girdling the main tree trunk will
have on the tree as a whole.

....

....

....

....

[Total 3 marks]

Score: ⬚

9

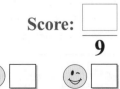

Plant Hormones

1 Plants increase their chances of survival if they can respond to environmental changes.
Complete the following passage about plant growth responses.

Plant shoots grow towards .... to maximise photosynthesis.

This growth response is known as positive ....

Plants can also sense gravity so their shoots and roots grow in the right direction.

This response is called .... Positive hydrotropism is a

growth response where plant roots grow towards ....

[Total 4 marks]

2 A student placed some germinating beans on the surface of some damp soil and left them for
five days. The appearance of the beans before and after the five day period is shown below.

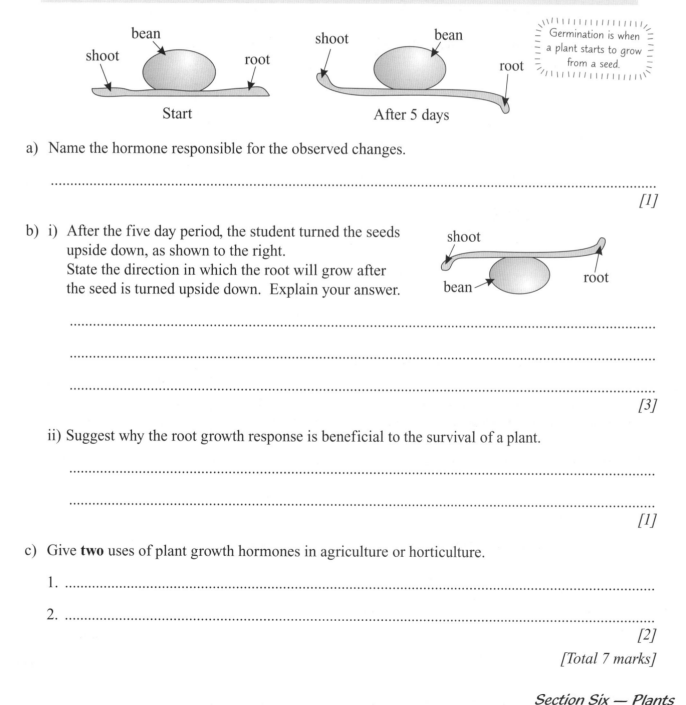

a) Name the hormone responsible for the observed changes.

....

[1]

b) i) After the five day period, the student turned the seeds
upside down, as shown to the right.
State the direction in which the root will grow after
the seed is turned upside down. Explain your answer.

....

....

....

[3]

ii) Suggest why the root growth response is beneficial to the survival of a plant.

....

....

[1]

c) Give **two** uses of plant growth hormones in agriculture or horticulture.

1. ....

2. ....

[2]

[Total 7 marks]

3 A teacher used three seedlings to investigate plant growth. Each seedling was prepared differently (see table). All three were placed in the same conditions, exposed to light from one direction and left for five hours. She recorded her results in the table below.

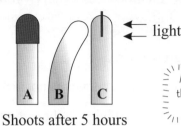

Seedling	Preparation	Observation after 5 hours
A	foil covering tip	no change
B	no preparation	tip bent towards light
C	mica strip through centre of tip	no change

Suggest why seedling **A** and seedling **C** failed to respond to the light.

Seedling **A** ....

....

Seedling **C** ....

....

[Total 4 marks]

4 A scientist is investigating the effect of auxin concentration on the growth of the roots in some identical plant cuttings. His measurements are shown in the table.

Plant cuttings are pieces of plants that have been cut off to grow into new plants.

Concentration of auxin (parts per million)	0	0.001	0.01	0.1	1
Length of root at start of investigation (mm)	20	20	20	20	20
Length of root 1 week after investigation started (mm)	26	32	28	26	21

a) What do the results suggest is the best concentration of auxin to use to encourage root growth?

....

[1]

b) From these results, describe the effect of high auxin concentration on the rate of root growth.

....

....

[2]

[Total 3 marks]

Score: ____

18

Section Six — Plants

Sexual Reproduction in Plants

1   The process of sexual reproduction in plants can be broken down into four main stages.

The four main stages are listed below. They are not listed in the correct order.

A | Transfer of the male gametes to the female ovules (pollination).

B | Ovules grow into seeds inside a fruit.

C | Fertilisation of ovules within the ovary.

D | The production of male and female gametes.

State the correct order of the stages shown above, starting with the earliest.

....

[Total 2 marks]

2   Sexual reproduction in plants is complex and involves many individual steps.

a) i) Complete the following passage about fertilisation in plants.

Pollen grains are produced by anthers and contain the male ....

Pollen grains attach to the .... on top of the female carpel. A pollen grain

produces a pollen tube, which grows through the .... and into the ovule.

[3]

ii) Describe the events that take place after the pollen tube has developed,
which lead to seed production.

....

....

....

....

[4]

b) Name the parts of the flower labelled **X**, **Y** and **Z** on the diagram.

i) **X** ....

ii) **Y** ....

iii)**Z** ....

[3]

[Total 10 marks]

Exam Practice Tip

You need to be able to recognise all of the structures in a flower that are involved in reproduction —
so make sure you learn all of their names <u>and</u> what they look like when you're learning about their roles.
Plant sexual reproduction looks pretty complicated, but if you break it down into steps it's easier to learn.

Score

12

Variation

1 Two sisters are identical twins.

a) One twin weighs 7 kg more than the other.
   Explain whether this is due to genes, environmental factors or both.

   ....

   ....
   [2]

b) One twin has a birthmark on her shoulder. The other does not.
   State whether birthmarks are caused by genes and explain your answer.

   ....
   [1]

   [Total 3 marks]

2 A racehorse owner wants to produce a successful foal from one of his thoroughbred mares.
He takes the mare to breed with a prize-winning stallion.

a) Explain why the foal will have similar characteristics to both its parents.

   ....

   ....

   ....
   [2]

b) Suggest whether or not the racehorse owner can confidently expect the foal to be a successful
   racehorse. Explain your answer.

   ....

   ....

   ....

   ....

   ....
   [3]

   [Total 5 marks]

   Score: ☐

   8

☹ ☐   ☺ ☐   ☺ ☐

The Structure of DNA

1 DNA in plants and animals is found in the form of chromosomes.

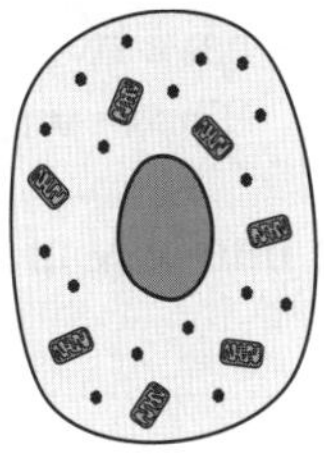

a) The diagram shows an animal cell. On the diagram, label where chromosomes are found with the letter **X**.

[1]

b) In the cells of some types of wheat, there are four of each chromosome present. How many of each chromosome are typically present in a human body cell?

....
[1]

c) Chromosomes are made from large molecules of DNA.

i) What do the letters DNA stand for? ....
[1]

ii) What is the role of DNA?

....

....
[1]

iii) Describe the structure of a DNA molecule.

....

....
[2]

[Total 6 marks]

2 The protein p53 is involved in controlling cell division and preventing the growth of tumours. In humans, the p53 protein is coded for by the TP53 gene.

a) Explain how the order of bases in a gene allows the production of a particular protein.

....

....

....

....
[3]

b) A mutation can change the order of bases in a gene. Suggest and explain what the consequences might be if the order of bases in the TP53 gene was changed.

....

....
[2] **Score:**

[Total 5 marks] **11**

X and Y Chromosomes

1 Chickens have two sex chromosomes, Z and W. In chickens, those with two Z chromosomes are male. Those with one Z chromosome and one W chromosome are female.

a) i) Give **one** way in which sex determination is **similar** in chickens and in humans.

....

[1]

ii) Give **two** ways in which sex determination is **different** in chickens and in humans.

1. ....

....

2. ....

....

[2]

b) Chickens have 39 pairs of chromosomes.
How many pairs of chromosomes do humans have? ....

[1]

[Total 4 marks]

2 Colour blindness in humans is caused by a recessive allele located on the X chromosome.
It is more common in men because men carry only one X chromosome.

A man who is colour blind has a child with a woman who does not have the recessive allele.

a) Draw a genetic diagram to show the inheritance of sex chromosomes in humans.

To get full marks for this question, you need to show the sex chromosomes of the parents and gametes, the possible combinations of sex chromosomes in the offspring and the sex of the offspring.

[4]

b) The couple's child is a boy.
Explain why the boy will not inherit colour blindness from his father.

....

....

....

[1]   **Score:**

[Total 5 marks]

9

The Work of Mendel

1   Mendel carried out experiments using pea plants and developed the
idea that plant characteristics were determined by 'inherited factors'.

a) i) Explain what Mendel meant by the term 'inherited factors'.

....

....

[2]

ii) Mendel said that some 'inherited factors' were dominant and some were recessive.
If an organism has both the dominant and the recessive 'inherited factors' for a characteristic,
which one is expressed?

....

[1]

Mendel crossed different combinations of tall and dwarf pea plants to investigate how tall and
dwarf characteristics were inherited. He discovered that the allele for tall plants (**T**) is dominant
and that the allele for dwarf plants (**t**) is recessive.

b) Draw a genetic diagram to show the outcome of crossing a tall pea plant (alleles, **TT**) with
a dwarf pea plant (alleles, **tt**).

Here's part of a genetic diagram to start you off:

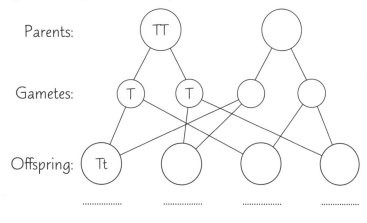

[3]

c) i) Draw a genetic diagram to show the outcome of crossing two tall pea plants,
both with the alleles **Tt**.

[3]

ii) In the cross you drew for part c) i), what is the
probability of one of the offspring plants being tall? ....

[1]   **Score:**

[Total 10 marks]   **10**

Genetic Diagrams

1 A woman carries a recessive allele, **b**, for red hair and a dominant allele, **B**, for brown hair.

a) Explain what is meant by the term *allele*.

....

[1]

b) The woman's genotype is **Bb**.

i) What is the term used to describe this genotype? ....

[1]

ii) Give the woman's phenotype. ....

[1]

[Total 3 marks]

2 Fruit flies usually have red eyes. However, there are a small number of white-eyed fruit flies. Having white eyes is a recessive characteristic.

Two fruit flies with red eyes have the heterozygous genotype for this characteristic. They are crossed to produce offspring.

a) Draw a genetic diagram to show the outcome of this cross.

Use **R** to represent the dominant allele and **r** to represent the recessive allele.

> You're told that the parents' genotypes are heterozygous — so they must be Rr.

[4]

b) i) State the probability that one of the fruit flies' offspring will have white eyes.

....

[1]

ii) The fruit flies have 60 offspring.
Calculate how many of the offspring are likely to have red eyes.

Answer ....

[2]

[Total 7 marks]

Exam Practice Tip

When you draw a genetic diagram, don't be tempted to take a short cut and miss out any stages — always show the parents' genotypes, the gametes' genotypes and the genotypes and phenotypes of the offspring. If you're not told how to construct your diagram, you can draw it however you like, as long as it's clear.

Score

10

More Genetic Diagrams

1 Albinism is a condition characterised by the lack of pigment in the hair and skin.
It is caused by the recessive allele **a**. The dominant allele **A** results in normal pigmentation.

A rabbit with albinism mated with a rabbit with no symptoms of the condition.
Below is a genetic diagram of the cross. A few details have been left out.

a) The parent with albinism has the genotype **aa**.
What term is used to describe this genotype?

....
[1]

b) Give the genotype of the parent without albinism
and the gametes this parent produces.

Parent genotype: .... Gametes genotypes: ....
[2]

c) i) From the genetic diagram, what percentage
of offspring are likely to have albinism? ....

[1]

ii) In reality, 60% of the offspring of these two parents had albinism.
Explain why this percentage is not the same as that suggested by the genetic diagram.

....
[1]
[Total 5 marks]

2 Huntington's disease is
transmitted by the dominant
allele **H**. The corresponding
recessive allele is **h**. This family
pedigree shows a family with a
history of this genetic disorder.

Key:
☐ Unaffected male
○ Unaffected female
■ Affected male
● Affected female

Alan   Amy
Brian   Carol   Ben   Chloe   Brook

Using the information given above, state what Amy's genotype must be. Explain your answer.

....
....
[Total 2 marks]

3 A gardener is breeding flowering plants for show. She has two of the same plant,
one with red flowers and one with white flowers. She knows that the allele for the
colour red (**R**) is dominant over the allele for the colour white (**r**). Suggest how she can
find out whether the plant with red flowers is thoroughbred red (**RR**) or hybrid red (**Rr**).

....
....
....

Score: ☐

[Total 3 marks]   **10**

Genetic Disorders

1 Down's Syndrome is an inherited genetic disorder that is not controlled by a single allele.

What is the cause of this disorder?

....

[Total 1 mark]

2 Polydactyly is a genetic disorder, which is caused by a dominant allele (**D**). The genotypes of two couples are shown in the diagram below.

Couple **A**

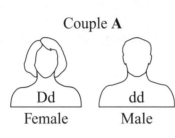

Dd — Female   dd — Male

Couple **B**

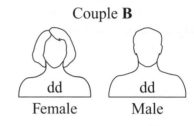

dd — Female   dd — Male

a) What are the symptoms of polydactyly?

....

[1]

b) Using the information in the diagram above, identify the individual that has polydactyly.

....

[1]

c) i) Construct a genetic diagram to show the potential phenotypes of the offspring from Couple **A**. Use the symbols **D** and **d** to represent the alleles.

[4]

ii) What is the probability of Couple A having a child with the disorder?   ....

[1]

[Total 7 marks]

3 Cystic fibrosis is a genetic disorder which affects cell membranes.
It is caused by a recessive allele. A man is a carrier of cystic fibrosis.
His wife is not a carrier of cystic fibrosis and does not have the disease.

a) The man and his wife are planning a family.

Wife's alleles

i) Complete the genetic diagram
on the right to show the cystic fibrosis
alleles that their offspring might inherit.

Use the symbols **F** and **f**
to represent the alleles.

Husband's
alleles

[2]

ii) What is the probability that the couple's child will have cystic fibrosis?

....
[1]

b) The man's brother has cystic fibrosis. His brother's wife does not have the disease and is not a
carrier for it. Could a child of theirs have cystic fibrosis? Explain your answer.

....

....
[2]
[Total 5 marks]

4 Sickle-cell anaemia is a genetic disorder caused by a recessive allele.
The recessive sickle-cell allele can be represented by the symbol **a**.
The dominant healthy allele can be represented by the symbol **A**.

a) Complete the following sentence about sickle-cell anaemia.

Sickle-cell anaemia is a disorder that affects the .... cells.
[1]

b) Describe how being heterozygous for the sickle-cell allele can be an advantage for an individual.

....

....
[1]

c) A genetic diagram to show the inheritance of sickle-cell alleles
from two heterozygous individuals is shown on the right.

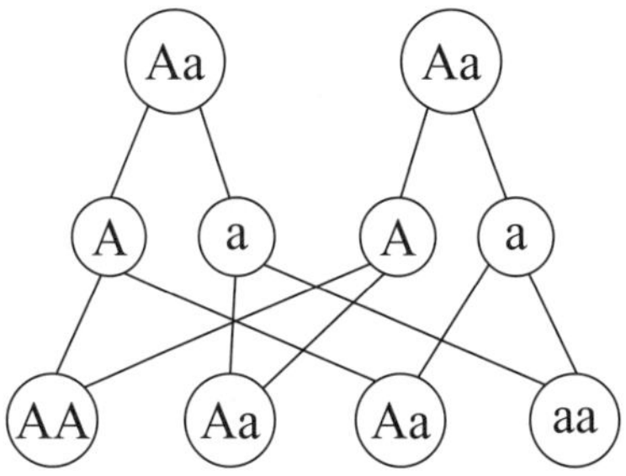

i) Give the genotype of carriers of this disorder. ....
[1]

ii) What is the percentage likelihood that the offspring of this
couple will have sickle-cell anaemia?

....
[1]
[Total 4 marks]

Aa   Aa

A a A a

AA Aa Aa aa

Score: ☐

17

Embryo Screening

1   During embryo screening, a cell is removed from an embryo and screened
to see if it contains alleles that cause certain genetic disorders.

a)   Before screening can take place, what must be extracted from the embryo's cell?

....
[1]

b)   Embryo screening involves the use of gene probes. Describe how a gene probe with a fluorescent
chemical attached can be used to locate an allele linked to a specific genetic disorder.

....

....

....

....
[3]

c)   *In this question you will be assessed on the quality of your English,
the organisation of your ideas and your use of appropriate specialist vocabulary.*

Embryos produced by IVF are screened for faulty alleles before being selected for implantation
into the mother's womb. Embryos conceived naturally can be screened while in the womb — if a
faulty allele is found, a decision has to be made about whether or not to terminate the pregnancy

Using the information above and your own knowledge, describe the arguments for and against
the screening of embryos.

....

....

....

....

....

....

....

....

....

....

....
[6]

[Total 10 marks]

Score:

10

Reproduction

1   Some organisms, such as humans, create new offspring
through the process of sexual reproduction.

   a) Choose words to complete the following sentence about sexual reproduction.

     Sexual reproduction is what happens when ....

     .... to form a single cell with the full number of chromosomes.

                                                 [2]

   b) Explain how the process of sexual reproduction leads to genetic variation in the offspring.

     ....

     ....

                                                 [1]

                                  [Total 3 marks]

2   A farmer grows potatoes. He looks for varieties of potato that have favourable characteristics,
such as good pest resistance and taste. Potato plants grow tubers underground, which are
genetically identical to the parent plant and can be used as 'seeds' to grow new plants from.

   a) What name is given to genetically identical individuals?

     ....

                                                 [1]

   b) What type of reproduction is involved in the development of tubers? Explain your answer.

     ....

     ....

     ....

                                                 [3]

   c) Suggest **one** benefit to the farmer of growing new potato plants from tubers.

     ....

     ....

                                                  [1]

                                  [Total 5 marks]

Exam Practice Tip

Knowing the difference between sexual and asexual reproduction is important. Both of these terms could
pop up anywhere in an exam, so make sure you've no problem with describing what each one means.
You need to remember that one produces genetically varied offspring and the other does not.

Score

8

Cloning

1 Plants can be cloned by tissue culture.

   a) Describe briefly how plant tissue culture works.

   ....

   ....

   [1]

   b) Tissue culture is commonly used by commercial plant growers to mass-produce plants.

   Suggest and explain **one** advantage of growing plants on a large scale using tissue culture, instead of allowing the plants to reproduce sexually.

   ....

   ....

   ....

   [2]

   [Total 3 marks]

2 A farmer wants to use embryo transplants to increase the size of her herd of beef cows. She begins by using sperm from her pedigree bull to artificially fertilise an egg from one of her best cows. This creates an embryo.

   a) Explain how the farmer will be able to produce several cloned calves from this one embryo.

   ....

   ....

   ....

   ....

   [3]

   b) The calves produced in this way are clones of each other, but not of the original cow and bull.

   i) Explain why the cloned calves are not genetically identical to the original cow and bull.

   ....

   ....

   [2]

   ii) Suggest an advantage to the farmer of increasing the size of her herd using embryo transplants, rather than allowing her animals to breed naturally.

   ....

   ....

   [1]

   [Total 6 marks]

3 The diagram shows the process of adult cell cloning, which can be used to clone a sheep.

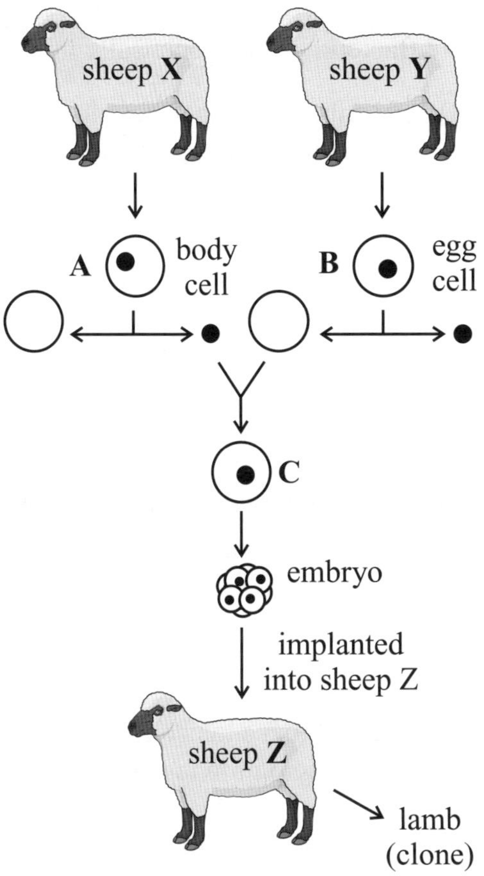

a) Which of the cells labelled **A**, **B** and **C** in the diagram are genetically identical to each other?
Explain your answer.

....

....

....
[2]

b) Which of the sheep (**X**, **Y** or **Z**) will the cloned lamb be genetically identical to?
Give a reason for your choice.

....

....

....
[2]

c) Once cell **C** has been formed, what stimulates this cell to divide and form embryo cells?

....
[1]

d) In normal sexual reproduction, fertilisation is the fusion of the female and male gametes and their genetic material. Explain the difference between this process and cloning.

....

....

....
[2]
[Total 7 marks]

4 Like many other scientific techniques, cloning animals has both pros and cons.

a) Suggest **one** way in which cloning animals could benefit society.

....
[1]

b) Suggest why the appearance of a new disease could be a problem for a population of cloned animals.

....

....
[2]   **Score:** ☐
[Total 3 marks]   **19**

Genetic Engineering

1 Bacteria can be genetically modified to produce large amounts of human insulin.

a) Describe how genetic engineering could be used to produce a bacterium capable of producing human insulin.

....

....

....

....

[3]

b) Plasmids are a type of vector used in genetic engineering. Name **one** other type of vector.

....

[1]

[Total 4 marks]

2 A US farmer's maize crop is being badly affected by an insect pest. He is considering switching to growing genetically modified (GM) maize, which is resistant to this pest.

a) i) Complete the following sentence:

The gene for pest resistance is inserted into the maize plants in the ....

stages of the plants' development, so that they will develop with the desired characteristic.

[1]

ii) Give **one** other way that a crop can be genetically modified to benefit farmers.

....

[1]

b) Explain how switching to the GM maize could benefit the farmer.

....

....

[1]

c) The farmer is unsure about growing a GM crop on his farm, as he has heard that some people are strongly against their use. Describe **two** concerns associated with the use of GM crops.

1. ....

2. ....

[2]

d) When GM crops are being developed they undergo extensive trials.
Suggest why it is important during these trials to grow the GM crops
in the same environmental conditions as conventional crops.

....

[1]   **Score:**

[Total 6 marks]   **10**

Adaptations

1   Some microorganisms live in deep-sea volcanic vents.

Use words from the box to complete the following passage.

| temperature | light levels | extremophobes | oxygen levels | pressure | extremophiles |

Microorganisms that have adapted to live in extreme conditions are known

as .... For example, bacteria living on deep-sea volcanic vents

can cope with very high .... and ....

[Total 3 marks]

2   The picture below shows a flea.

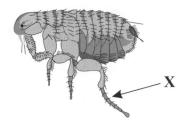

X

a) Where do fleas typically live?

....

[1]

b) Complete these sentences about the flea.

You'll need to write more than one word to fill in each of these gaps.

The flea has sharp mouthparts to ....

The flea has a flattened body so that ....

....

The body is hard so .... when the

animal scratches at it.

[3]

c) Give the function of the structure marked **X** on the diagram.

....

[1]

[Total 5 marks]

3 Tapeworms are parasites that live inside the intestines of mammals and some other animals.

a) How are tapeworms adapted to fix themselves to the wall of the intestine?

....

[1]

b) What is the main function of the segments in a tapeworm's body?

....

[1]

c) State and explain **two** other adaptations tapeworms have developed for living inside the intestines.

1. ....

....

2. ....

....

[4]

[Total 6 marks]

4 Black bears live in North America, mostly in forests. They have a varied diet and feed on insects, nuts, berries and other plant material, as well as other mammals and fish. They have five toes on each foot with long curved claws.

a) Suggest **two** advantages to the bears of having these long claws.

1. ....

....

....

2. ....

....

....

> Use the information you've been given in the question intro and in the photo to help you here.

[2]

b) i) Black bears often hibernate in winter. This means they lower their metabolism during the winter months. Suggest and explain **one** advantage of hibernating during the winter.

....

....

[2]

ii) There is evidence that black bears in some areas travel long distances to find good food sources in late summer. Suggest a reason for this behaviour.

....

....

[1] Score:

[Total 5 marks]

19

The Malaria Parasite

1   The parasite that causes malaria in humans has several different forms.

a) When a mosquito sucks blood from an animal with malaria, it takes up a form of the parasite
called a gametocyte. Describe how gametocytes develop into sporozoites in the mosquito's body.

....

....
      [2]

b) *In this question you will be assessed on the quality of your English,
the organisation of your ideas and your use of appropriate specialist vocabulary.*

When the mosquito bites a human, it transfers sporozoites into the person's bloodstream.

Describe the stages of the malaria parasite's life cycle that take place in the human body.

....

....

....

....

....

....

....

....

....

....

....

....

....
      [6]

c) Each form of the malaria parasite has a different structure and appearance.
Suggest a reason for this.

....
      [1]

      [Total 9 marks]

Exam Practice Tip

There are a lot of long, difficult-looking words to do with the malaria parasite, but you need to learn them all
for your exams. If a question comes up where you're told that you'll be marked on your use of scientific terms,
it's important to make sure you get all the relevant words in. You need to try to spell them all correctly too.

Score

9

Competition and Environmental Change

1   Animals and plants require resources from their surroundings to survive.

a) What would happen if two species in a habitat need the same resource?

....

[1]

b) Give **one** way that organisms are dependent on other species for their survival.

....

[1]

c) i) Suggest **two** resources that animals might need from their surroundings.

1. ....

2. ....

[2]

ii) Suggest **two** resources that plants might need from their surroundings.

1. ....

2. ....

[2]

[Total 6 marks]

2   The graph shows the maximum height up a mountain at which a snail species was found between 1916 and 2008.

a) From the graph, briefly describe the change in the distribution of the snail species from 1916 to 2008.

....

....

....
[1]

During the time period shown on the graph, the average temperature on the mountain increased.

b) Use this information to suggest how the change in temperature may have affected the distribution of the snail species.

....

[1]

c) Other than temperature, give **one** example of a
non-living factor that can affect the distribution of a species. ....

[1]

d) Give another example of an organism whose
distribution has been affected by a change in its environment. ....

[1]

[Total 4 marks]

3 Algae are tiny, plant-like organisms that make their own food through photosynthesis. They are eaten by fish. The graph shows how the size of a population of algae in a pond varied in one year.

a) Suggest **two** conditions that may have changed in the pond to give more algae in April than in January.

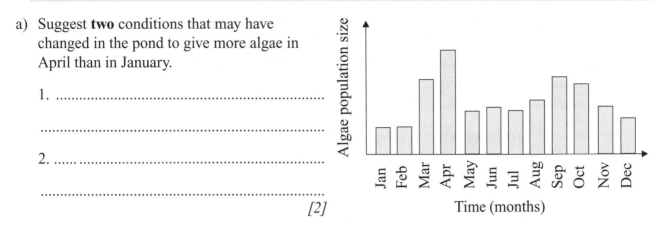

1. ....

....

2. ....

....

[2]

b) The number of fish in the pond increased rapidly during one month of the year. Suggest which month this was. Explain your answer.

....

....

....

[3]

[Total 5 marks]

4 The graph shows how the populations of rabbits and foxes in a community changed over time.

Population size

Key:
------- Rabbits
——— Foxes

Time

Think about how changes in the population of rabbits might affect the population of foxes, and vice versa.

a) Describe and explain the relationship between the population sizes of these two species.

....

....

....

....

....

[4]

b) Apart from food, suggest **one** factor that the foxes in this community may be competing for.

....

[1]   **Score:**

[Total 5 marks]   **20**

Section Eight — Adaptation and Environment

Measuring Environmental Change

1 Mayfly larvae and sludgeworms are invertebrates. They can be used to study water pollution.

Water samples were taken from a river at three different distances from a sewage outlet.
The number of mayfly larvae and sludgeworms in each sample was recorded.
The results are shown in the table.

Distance (km)	No. of mayfly larvae	No. of sludgeworms
1	3	20
2	11	14
3	23	7

a) Draw a conclusion about the relationship between each of these organisms
   and the level of pollution in the water.

   ....

   ....
   [2]

b) The concentration of dissolved oxygen in each water sample was also measured.
   For which sample would you expect the oxygen concentration to be highest?

   ....
   [1]
   [Total 3 marks]

2 A group of scientists investigated the local air quality. Data was collected on the sulfur dioxide
concentration in the air at different sites in a town. The data is shown in the table below.

Site	1	2	3
Sulfur dioxide concentration (micrograms/m^3)	9.8	9.4	7.1

The sulfur dioxide concentration was measured directly, using specialist equipment.

a) Give **one** example of a living indicator that could
   have been used to monitor air quality in this study. ....
   [1]

b) Evaluate the use of specialist equipment to measure air pollution in this study,
   rather than a living indicator.

   ....

   ....

   ....

   ....

   ....
   [3] **Score:**
   [Total 4 marks]   **7**

Monitoring the Distribution of Organisms

1 A student wanted to investigate the number of daisies on the school field. He placed a 1 m^2 quadrat down at eight random points in the field and counted the number of daisies in each quadrat. He recorded his results in the table below.

Quadrat	1	2	3	4	5	6	7	8
Number of daisies	3	1	2	1	4	3	0	2

a) Suggest **one** way that the student could have ensured the quadrats were placed at random locations.

....

....

[1]

b) i) Calculate the mean number of daisies per quadrat in the school field.

Mean number of daisies = ... per quadrat

[2]

ii) Calculate the median number of daisies per quadrat in the school field.

Median number of daisies = ... per quadrat

[2]

iii) The total area of the field is 5600 m^2.
Use your answer to b i) to estimate the total population of daisies in the field.

Total population of daisies = ...

[2]

c) Suggest how the reproducibility of this investigation could be improved.

....

[1]

[Total 8 marks]

2   A student investigated the distribution of dandelions across
      a field next to a wood. A sketch of the area is shown below.

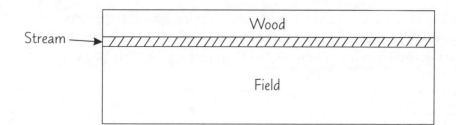

The following table shows the results of the investigation.

Number of dandelions per m²	5	9	14	19	26
Distance from wood (m)	2	4	6	8	10

a) Describe a procedure the student could have used to obtain these results.

....

....

....

....
                                                                                                                                         [3]

b) Describe the trend in the results.

....

....
                                                                                                                                         [1]

The student wanted to see if the trend in the distribution of dandelions was due to varying light
intensity across the area investigated. He measured the light intensity and found it decreased
towards the wood. He concluded that the distribution of dandelions depends on light intensity.

c) Explain why the results are not valid.

> Remember: valid results are
> repeatable, reproducible and
> answer the original question.

....

....

....
                                                                                                                                         [3]

d) Suggest another variable that could have affected the distribution of dandelions.

....
                                                                                                                                         [1]
                                                                                                                            [Total 8 marks]

Exam Practice Tip   **Score**

You need to know the differences between the three different types of average — mean, median and mode.
Make sure you know how to work them all out and, if you get a question on averages, make sure you know
which you are being asked for before you start your answer. Always remember to show your working too.
                                                                                                                                  16

Human Impact on the Environment

1   The graph shows how the human population of the Earth has changed over the last 1000 years.

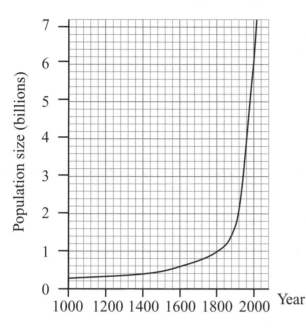

a) Describe the trend in population size over the last 1000 years, as shown by the graph.

Population slowly increased up to 1 billion in 1800. Then the population very rapidly increased to over 7 billion in the year 2000

[2]

b) The increase in the human population has caused a reduction in the amount of land available for other animals and plants. Suggest **three** reasons why.

1. *Building*

2. *Quarrying*

3. *Farming*

[3]

c) As the human population has grown, the amount of waste it produces has increased. Give **two** types of waste produced by humans and state how each one pollutes the environment.

1. *Sewage, which causes eutrophication*

2. *Smoke, causes air pollution*

[4]

d) Other than the increase in human population, suggest **one** reason why the amount of waste humans produce has increased.

Standard of living has increased, meaning humans consume more

[1]

[Total 10 marks]

Score:

10

Eutrophication

1   A river flows through some farmland. Data showing the concentration of nitrates in the river and the average number of fish per cubic metre of water between 2002 and 2010 are shown below.

	2002	2004	2006	2008	2010
Nitrate concentration (mg per litre)	22	33	48	63	74
Average number of fish per m³	23	21	10	1	0

a) Describe the relationship between the nitrate concentration and the average number of fish per cubic metre between 2002 and 2010.

As nitrate concentration increases, the average number of fish per m³ decreases
[1]

b) Suggest an explanation for the trend in the nitrate concentration over this time period.

more fertilisers are being used
[1]

c) Explain how eutrophication could have caused the observed change in the average number of fish per cubic metre of water.

More fertilisers get washed into lakes/rivers which makes plants and algae grow faster. They then use up space so other organisms die because they can't photosynthesise. Microorganisms that feed off of dead organisms reproduce and deplete all the oxygen from respiring. Therefore aerobic organisms e.g. fish die because there is no oxygen for them to respire
[5]

[Total 7 marks]

2   An investigation was carried out into the number of microorganisms along a stream. A sewage outflow pipe was located mid-way along the study site. The results are shown to the right.

Describe and explain the change in number of microorganisms downstream from the outflow pipe.

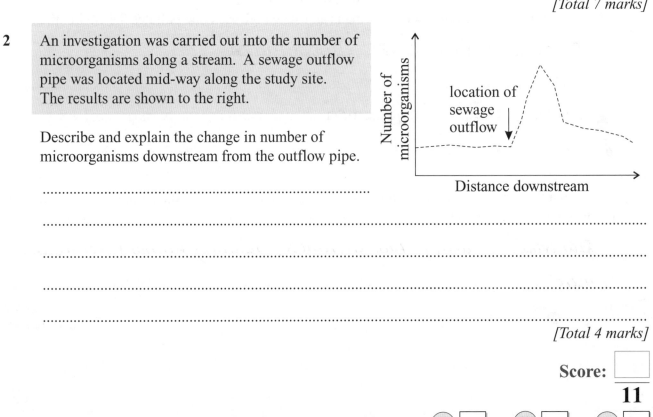

....

....

....

....

....

[Total 4 marks]

Score:

11

Deforestation and Destruction of Peat Bogs

1 Deforestation of tropical rainforests is causing an increasing number of problems.

a) Give **two** reasons why deforestation is occurring in tropical areas.

1. ....

2. ....

[2]

b) i) Explain how deforestation affects the amount of carbon dioxide removed from the atmosphere.

....

....

[2]

ii) Give **two** reasons why deforestation increases the amount of carbon dioxide released into the atmosphere.

1. ....

2. ....

[2]

c) Deforestation may reduce the *biodiversity* of a tropical rainforest.
What is meant by the term *biodiversity*?

....

[1]

[Total 7 marks]

2 The destruction of peat bogs releases carbon dioxide into the atmosphere.

a) Describe how a peat bog is formed.

....

....

[2]

b) Explain why draining a peat bog increases the level of carbon dioxide in the atmosphere.

....

....

[2]

c) Suggest **one** way a gardener could help reduce the number of peat bogs being drained.

....

[1]

[Total 5 marks]

Score:

12

Carbon Dioxide and the Greenhouse Effect

1 The concentration of carbon dioxide in the atmosphere has been increasing since the 1800s.

a) i) Describe and explain the effect of this increase in carbon dioxide on the Earth's temperature.

....

....

....
[2]

ii) Name **one** other gas that affects the Earth's temperature in the same way as carbon dioxide.

....
[1]

b) Carbon dioxide can be sequestered in forests.

i) Why is it important that carbon dioxide is sequestered in forests?

....

....
[1]

ii) Give **two** other places where carbon dioxide can be sequestered.

1. .... 2. ....
[2]

[Total 6 marks]

2 Scientists have used model simulations to estimate how much the average global temperature will increase in the future. The scientists have produced data using a range of different scenarios. Some of their estimates are shown on the graph below.

a) Suggest which scenario (B1-A1FI) involves the largest amount of gases such as carbon dioxide being released into the atmosphere. Explain your answer.

....

....

....

....
[3]

b) Other than a rise in temperature, give **one** effect that scenarios A2 and A1FI may have on the Earth.

....
[1]

[Total 4 marks]

Score: ☐

10

Section Nine — Environment, Evolution and Energy Transfer

Climate Change

1 Sea turtles spend most of their lives at sea, but come ashore to lay their eggs on sandy beaches.

Suggest how the distribution of sea turtles is likely to be affected if the Earth's average temperature increases by a few degrees. Explain your answer.

....

....

....

[Total 3 marks]

2 Swallows migrate across parts of Europe and arrive in Britain in spring.

The graph shows two sets of data produced by different scientists studying the migration pattern of swallows.
Data set **1** was collected in a study that lasted 222 years. Data set **2** was collected in a study that lasted 43 years. In each study, the scientists recorded the first day in the year that a swallow was seen in Britain. These results have been plotted against the average April temperature in Britain that year.

a) Describe the overall trend shown on the graph.

....

....

[1]

b) Evaluate whether the results of the studies, as shown by the graph, provide valid evidence that the migration pattern of swallows has been affected by global warming.

....

....

....

....

[4]

[Total 5 marks]

Score:

8

Evolution

1 The diagram shows the evolutionary relationships between four different species.

Dolphins   Mice   Rays   Sharks

a) Which species on the diagram are **rays** most closely related to? Explain your answer.

....

....
[2]

b) Sharks and dolphins share similarities even though they are not closely related.
Suggest **one** thing this could tell us about their ecological relationship.

....
[1]

c) Evolutionary trees are created by studying the similarities and differences between different organisms. Give **one** other reason why scientists might study the similarities and differences between organisms.

....
[1]
[Total 4 marks]

2 Darwin's theory of evolution states that all organisms have evolved from simple life forms.

a) Complete the following sentence.

According to Darwin's theory, life on Earth began

more than .... years ago.
[1]

b) Ancient horses had shorter legs than modern horses.
A student suggests that horses may have evolved longer legs to run away from predators faster.
Based on the student's suggestion and the theory of evolution by natural selection,
describe how horses might have evolved to have longer legs.

....

....

....

....

....

....
[4]   **Score:**

[Total 5 marks]   **9**

More About Evolution

1 When Charles Darwin first published his theory of evolution many people did not accept it.

a) Give **three** reasons why Darwin's theory was not initially accepted.

1. ....

....

2. ....

....

3. ....

....
[3]

b) One alternative to Darwin's theory of evolution was put forward by Lamarck.

i) Complete the following sentences using words from the box.

developed	longer	stunted	shorter	similar	evolved	different

Lamarck argued that if a characteristic was used a lot by an organism then it would become

more .... during its lifetime. For example, if an anteater used its tongue

a lot to reach ants in anthills, its tongue would get .... The offspring of

this anteater would then inherit .... length tongues to their parents.
[3]

ii) Why has Lamarck's hypothesis been rejected as the main way that evolution occurs?

....
[1]
[Total 7 marks]

2 A pigeon hatches with an unusually long beak. While the pigeon is still young it loses its right leg in an accident, causing its left leg to grow stronger to support its weight. The pigeon breeds. Its offspring have unusually long beaks and both legs are of normal strength.

Does this support Lamarck's hypothesis about evolution? Explain your answer.

....

....

....
[Total 1 mark]

Score: ⬜

8

Fossils

1 Scientists are trying to discover more about prehistoric life by studying fossils.

a) What is a *fossil*?

....

....

[1]

b) The diagram shows a fossil of a dinosaur's footprint that the scientists are studying.

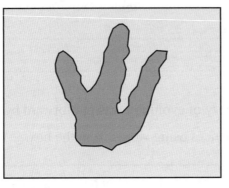

i) Describe how this fossil might have formed.

....

....

....

[2]

ii) The dinosaur footprint is compared to fossils of the bones in a dinosaur's foot, which were found in rock nearby. Describe how the fossils of the foot bones might have formed.

....

....

....

....

[2]

c) Despite the discovery of many fossils around the world, scientists are still uncertain about how life on Earth began. Explain why.

....

....

....

....

[3]

[Total 8 marks]

2 The diagram shows a chalk cliff face where fossils have been found.

a) Marked on the diagram are the places where three fossils were found in the cliff.

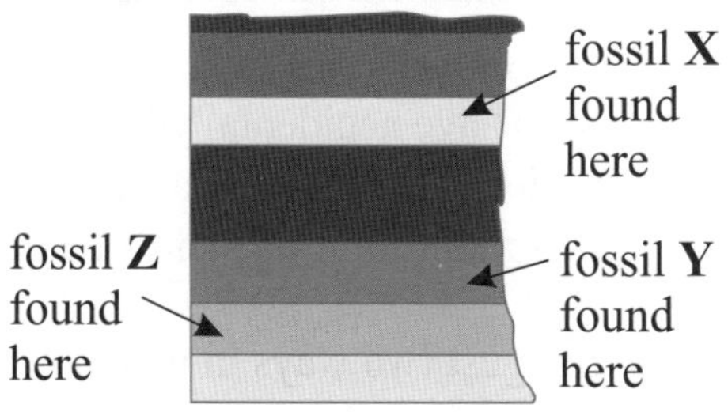

fossil **X** found here

fossil **Z** found here

fossil **Y** found here

i) Which fossil, **X**, **Y** or **Z**, is likely to be formed from the most recent organism? Explain your answer.

....

....
[1]

ii) Fossils **X**, **Y** and **Z** are all fossils of shelled sea creatures.
Suggest why there are more fossils of shelled sea creatures than of plants.

....

....

....
[2]

b) Further along the cliff, a group of fossils of one type of organism was found.
The fossils were sorted into age order, as shown in the diagram.

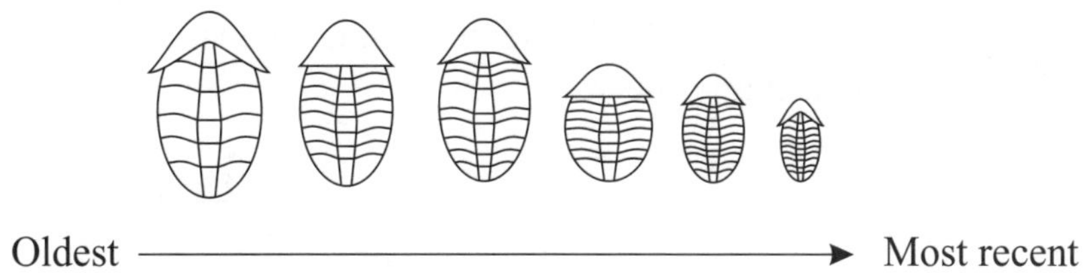

Oldest ⟶ Most recent

Suggest why the scientists put the fossils into age order when they were studying them.

....

....
[1]
[Total 4 marks]

3 In 1991 the fossilised body of a man was found, more than 5000 years after his death. The man was found in the Alps, buried in the ice.

Suggest how the man's body was preserved.

....

....

....
[Total 2 marks]

Score:

14

Section Nine — Environment, Evolution and Energy Transfer

Extinction and Speciation

1 Rats have accidentally been taken all over the world on ships by explorers and traders. In New Zealand, where there are no native rats, scientists are worried that rats from trade ships could cause the extinction of some species of bird.

a) Suggest **two** ways in which rats could cause the number of birds to decrease.

1. ....

2. ....

[2]

b) The extinction of a species is not always caused by other species.
Suggest **two** other reasons why a species may become extinct.

1. ....

2. ....

[2]

[Total 4 marks]

2 Two lakes, **A** and **B**, used to be connected by a river, which has now dried up. Each lake contains a different species of fish. A scientist studies the fish and concludes they evolved from a single species.

Lake A   **Lake B**

Dried up river

a) What is the name of the process by which new species develop?

....

[1]

b) Describe how the two new species of fish evolved from a single species after the river dried up.

....

....

....

....

....

....

....

[5]

[Total 6 marks]

Score:

10

Pyramids of Biomass

1 A single robin has a mass of 15 g and eats caterpillars. Each robin
eats 25 caterpillars that each have a mass of 2 g. The caterpillars
feed on 10 stinging nettles that together have a mass of 500 g.

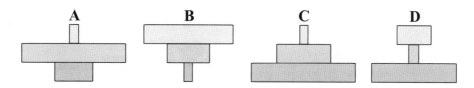

Which pyramid, **A**, **B**, **C** or **D**, represents a pyramid of biomass for the organisms
mentioned above? Explain your answer.

....

....

[Total 2 marks]

2 In the 1950s a chemical called DDT was used to control animal pests. DDT was later
discovered to be toxic and was detected at very high levels in the tissues of organisms
across food chains. The pyramid of biomass below shows the concentration of DDT
in the tissues of organisms at each stage in parts per million (ppm).

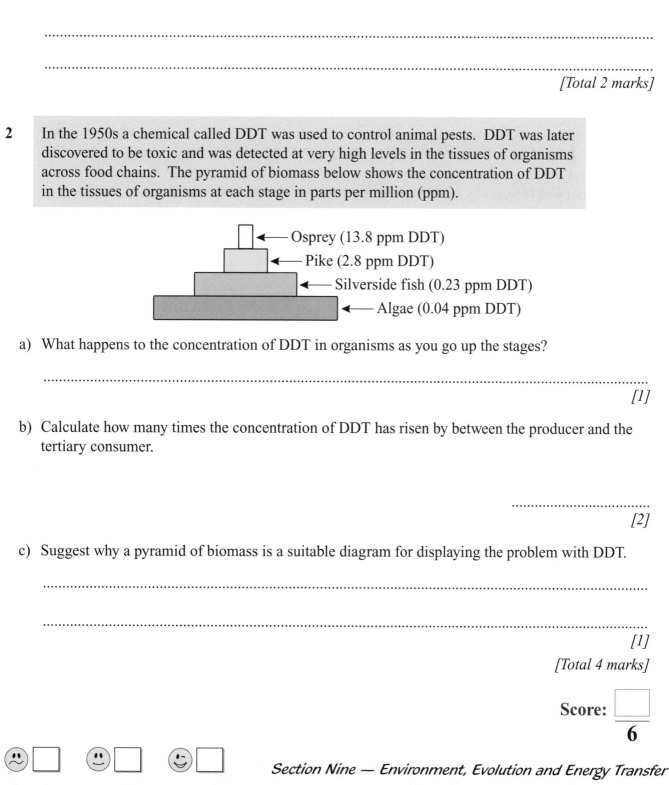

← Osprey (13.8 ppm DDT)
← Pike (2.8 ppm DDT)
← Silverside fish (0.23 ppm DDT)
← Algae (0.04 ppm DDT)

a) What happens to the concentration of DDT in organisms as you go up the stages?

....

[1]

b) Calculate how many times the concentration of DDT has risen by between the producer and the
tertiary consumer.

....

[2]

c) Suggest why a pyramid of biomass is a suitable diagram for displaying the problem with DDT.

....

....

[1]

[Total 4 marks]

Score:

6

Energy Transfer

1 The diagram shows how energy is transferred along a food chain.

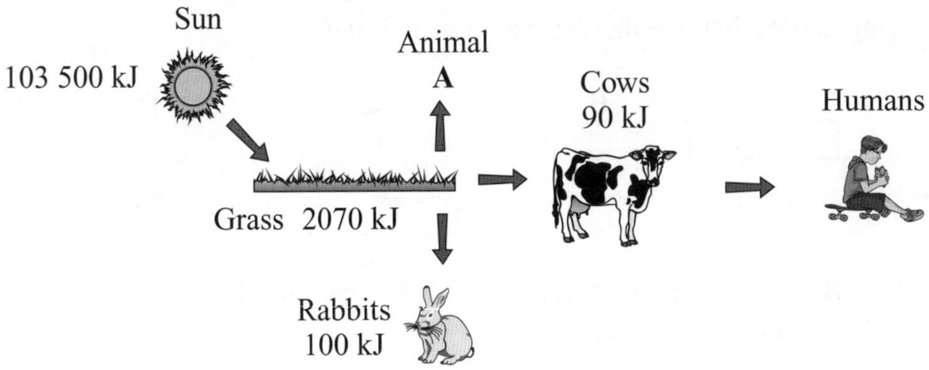

Sun
103 500 kJ

Animal
A

Cows
90 kJ

Humans

Grass 2070 kJ

Rabbits
100 kJ

a) Describe how energy is transferred from the Sun to a rabbit in this food chain.

....

....

....

[3]

b) Suggest **one** reason why so little of the Sun's energy is transferred to the grass.

....

....

[1]

c) The efficiency of energy transfer from the grass to the next stage is 10%.
   Calculate how much energy is available to animal **A**.

Tip: animal A is not the only
organism in the second stage
of the food chain.

Energy available to animal **A** = ... kJ

[2]

d) Explain how energy and biomass are lost between the rabbits and the cows.

....

....

....

....

[3]

[Total 9 marks]

2 Three different food chains are shown below.

A grass ➡ cattle ➡ human

B pondweed ➡ small fish ➡ salmon ➡ human

C wheat ➡ human

Which of the food chains, **A**, **B** or **C**, shows the most efficient food production for humans? ☐

Explain your answer.

....

....

[Total 2 marks]

3 Many turkey farmers increase the efficiency of meat production by controlling the conditions in which they keep their birds.

The table below lists some of the rearing conditions for 20 week old turkeys at three different turkey farms. All the turkey farms rear the same type of turkey.

	Turkey Farm		
	A	**B**	**C**
Space per bird (m²)	0.55	0.85	0.47
Temperature of turkey shed (°C)	17	20	22

a) Based on this information, suggest which turkey farm (**A**, **B** or **C**) produces turkey meat most efficiently. Explain your answer.

....

....

....

....

[3]

b) Give **one** benefit to the consumer of turkey meat being produced as efficiently as possible.

....

[1]

[Total 4 marks]

Score: ☐

15

Problems with Food Production and Distribution

1   In 2012, free range eggs made up roughly 45% of all the eggs that were
    packed and processed in the UK, and factory farmed eggs made up roughly
    50%. Free range eggs come from birds that are able to roam about freely.

a) Evaluate the pros and cons of producing factory farmed eggs compared to free range eggs.

....

....

....

....

....

[4]

b) Some of the eggs on sale in the UK have been imported from abroad.
    Explain why importing food from abroad can be bad for the environment.

....

....

....

[2]

[Total 6 marks]

2   Overfishing has become a global problem in recent years.

Overfishing happens when fish are caught quicker than they can breed.

a) Suggest **one** reason why the government is trying to reduce overfishing.

....

....

[1]

b) Describe and explain **two** ways in which overfishing could be reduced.

1. ....

....

2. ....

....

[4]

[Total 5 marks]

Exam Practice Tip

If you're asked to evaluate the pros and cons of different types of food production make sure you give a balanced argument. Also make it really clear which type of food production you're referring to in each of your statements — if the examiner's not convinced you really know what you're talking about you may not pick up marks.

Score

11

Decay and the Carbon Cycle

1   Carbon is constantly being recycled. The diagram shows
     some of the processes occurring in the carbon cycle.

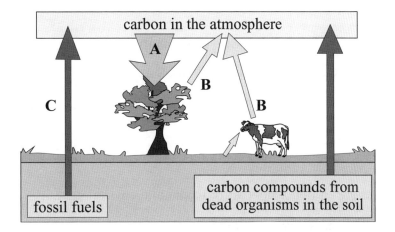

a) i) Name the process, labelled **A**, that removes carbon from the atmosphere.

....

[1]

ii) Name the gas in which carbon is removed from the atmosphere by process **A**.

....

[1]

b) i) Name the process, labelled **B**, by which plants and animals return carbon to the atmosphere.

....

[1]

ii) Describe how carbon moves from the atmosphere into animals such as cows.

....

....

[2]

c) i) Explain why fossil fuels contain carbon.

....

....

[1]

ii) Describe how the carbon from fossil fuels is released back into the atmosphere (process **C**).

....

....

[1]

[Total 7 marks]

2 Microorganisms break down dead animals and plants.

a) Describe the conditions in which microorganisms break down plants and animals most quickly.

....

....

[2]

b) Microorganisms respire as they break down dead plants and animals.

Name the gas that is released during this process.

....

[1]

[Total 3 marks]

3 Stable communities are an important part of the natural environment.

a) What is meant by the term *stable community*?

....

....

[1]

b) *In this question you will be assessed on the quality of your English, the organisation of your ideas and your use of appropriate specialist vocabulary.*

Grass needs both carbon dioxide and nitrogen to grow properly. When a lawn is cut, leaving the grass cuttings on the lawn may help to encourage a stable community. Explain how.

....

....

....

....

....

....

....

....

....

....

....

....

Score

....

....

[6]

[Total 7 marks] **17**

Candidate Surname		Candidate Forename(s)	

Centre Number		Candidate Number	

Level 1/2 Certificate in Biology
Paper 1

Practice Paper
Time allowed: 90 minutes

You must have:
- A ruler.
- A calculator.

Total marks:

Instructions to candidates
- Use **black** ink to write your answers.
- Write your name and other details in the spaces provided above.
- Answer **all** questions in the spaces provided.
- In calculations, show clearly how you worked out your answers.

Information for candidates
- The marks available are given in brackets at the end of each question.
- There are 90 marks available for this paper.
- You should answer Question 9 (a) with continuous prose.
  You will be assessed on the quality of your English,
  the organisation of your ideas and your use of
  appropriate specialist vocabulary.

Advice for candidates
- Read all the questions carefully.
- Write your answers as clearly and neatly as possible.
- Keep in mind how much time you have left.

Answer **all** questions

1   Cells are the building blocks of all organisms on the planet.

1 (a)   The diagram shows an animal cell.

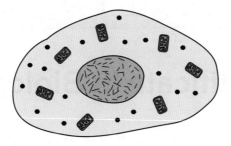

1 (a) (i) Give **two** ways in which this cell is different from a bacterial cell.

1. ....

....

2. ....

....
                                                                                                    [2]

1 (a) (ii) Name the structure in an animal cell where protein synthesis occurs.

....
                                                                                                    [1]

1 (b)   Describe the structure of a yeast cell.

....

....

....
                                                                                                    [3]

1 (c)   Plant cells are surrounded by a cell wall.

Name the substance that plant cell walls are made of and describe how this substance is produced.

....

....

....

....

[3]

1 (d)   The diagram below shows some molecules moving through a cell membrane.

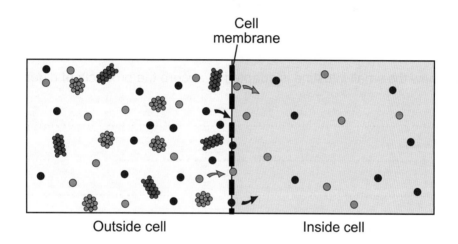

Describe and explain the process by which oxygen moves into cells.

....

....

....

....

[3]

[Total 12 marks]

Turn over for the next question

Turn over ▶

2   The diagram below shows nutrients being absorbed from the small intestine into the blood.

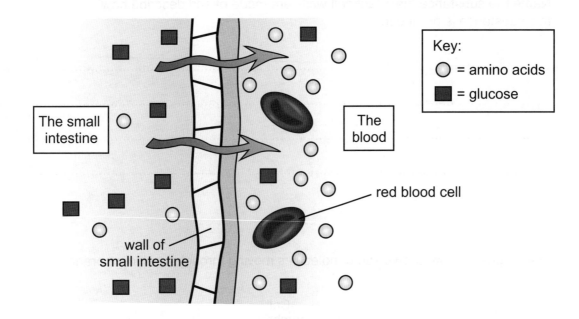

2 (a)   Explain how the small intestine is adapted to absorb the products of digestion.

....

....

....

....

....

[3]

2 (b)   The diagram shows glucose being absorbed into the blood by diffusion.
Describe how the amino acids are absorbed into the blood.

....

....

....

[2]

2 (c) The diagram shows red blood cells in the blood.

2 (c) (i) State the function of red blood cells.

....

[1]

2 (c) (ii) Name another type of blood cell and describe its function.

....

....

[2]

[Total 8 marks]

Turn over for the next question

Turn over ▶

106

3   Cystic fibrosis is a genetic disorder caused by a recessive allele.

A couple have a baby boy. The doctor tells them that the baby has inherited cystic fibrosis. Neither parent shows signs of the disorder.

3 (a) (i) In the space below, construct a diagram to show how the baby inherited cystic fibrosis.

Use **F** to represent the dominant allele and **f** to represent the recessive allele.

[4]

3 (a) (ii) Is the baby homozygous or heterozygous for this condition?
Explain your answer.

....

....
[1]

3 (b)   The doctor tells the parents that if they have another child, the embryo can be tested by embryonic screening to see if the child will have cystic fibrosis.

Explain how embryonic screening could be used to test the embryo for cystic fibrosis.

....

....

....

....
[4]
[Total 9 marks]

4   The peppered moth is an insect that lives on the trunks of trees in Britain. The moths are prey for birds such as thrushes.

The peppered moth exists in two varieties:

1. A light-coloured variety — they are better camouflaged on tree trunks in unpolluted areas.

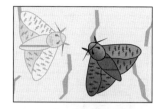

2. A dark-coloured variety — they are better camouflaged on sooty tree trunks in badly polluted areas.

The dark variety of the moth was first recorded in the north of England in 1848. It became increasingly common in polluted areas until the 1960s, when the number of soot-covered trees declined because of the introduction of new laws.

4 (a)   Using the idea of natural selection, explain why the dark variety of moth became more common in soot-polluted areas.

....

....

....

....

....

[4]

Question 4 continues on the next page

Turn over ▶

4 (b) The following bar charts show the percentages of dark- and light-coloured peppered moths in two different towns.

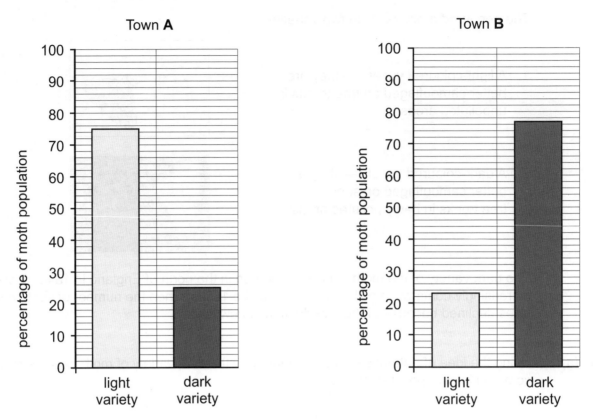

4 (b) (i) What does this information suggest about the relative levels of soot-pollution in towns **A** and **B**? Explain your answer.

....

....

[2]

4 (b) (ii) Suggest **one** problem with using peppered moths as an indicator of air pollution.

....

....

[1]

[Total 7 marks]

5   Glucose is a simple sugar that provides the body with energy.

5 (a)   Carbohydrates in the diet are broken down into simple sugars by enzymes such as amylase.
             Give **two** places in the digestive system where amylase acts.

             1. ....*Practice Paper*

             2. ....

                                                                                                                  [2]

5 (b)   When a person's blood glucose level is low their body responds to raise it.
             Describe how this happens.

             ....

             ....

             ....

             ....
                                                                                                                  [3]

5 (c)   The graph shows how two people's blood glucose level changes after they have eaten
             similar meals. One of these people has Type 1 diabetes.

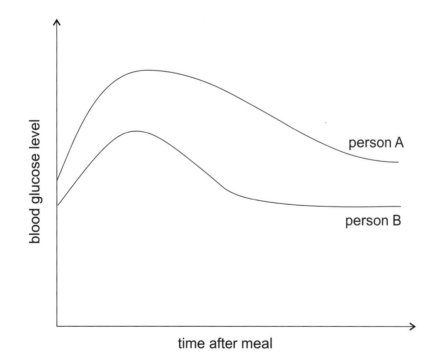

Question 5 continues on the next page

Turn over ▶

5 (c) (i) Which of the people shown in the graph is more likely to have Type 1 diabetes? Explain your answer.

....

....

....

....

....

[3]

5 (c) (ii) Give **two** ways that Type 1 diabetes can be controlled.

1. ....

....

2. ....

....

[2]

[Total 10 marks]

6 The graph below shows the blood water content of a person over a period of several hours.

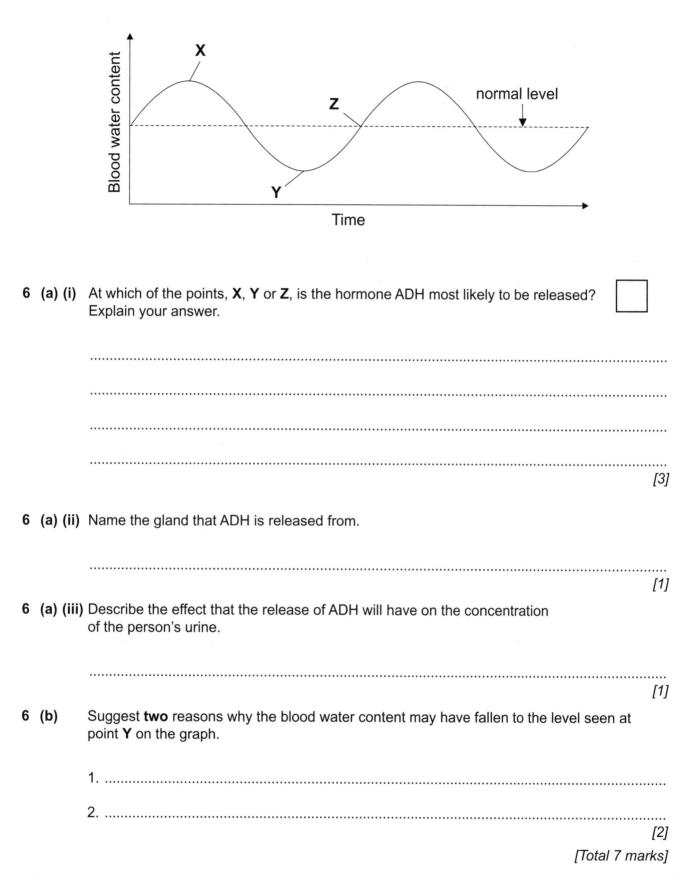

6 (a) (i) At which of the points, **X**, **Y** or **Z**, is the hormone ADH most likely to be released? Explain your answer.

☐

....

....

....

....

[3]

6 (a) (ii) Name the gland that ADH is released from.

....

[1]

6 (a) (iii) Describe the effect that the release of ADH will have on the concentration of the person's urine.

....

[1]

6 (b) Suggest **two** reasons why the blood water content may have fallen to the level seen at point **Y** on the graph.

1. ....

2. ....

[2]

[Total 7 marks]

Turn over for the next question

Turn over ▶

7   A crop plant has been genetically modified to make it resistant to a particular herbicide.

7 (a)   The GM crop plant contains a gene for herbicide resistance taken from another plant species.

     (a) (i)   Explain how genetic engineering would have been used to produce the GM crop plant.

....

....

....

....

                                                           [3]

7 (a) (ii)   The gene for herbicide resistance codes for a protein.
                 Explain how a gene can code for a protein.

....

....

                                                            [2]

The GM crop plant is already being grown by some farmers.

There are concerns that, as a result, wild grasses growing near to the crop plant might have also become resistant to herbicides. Some scientists carried out an investigation to discover whether this had happened.

The scientists sprayed herbicide onto 100 grass plants in an area next to the GM crop, and onto 100 grass plants from a second area 2 km away from the GM crop.

Their results are shown in the table below.

Number of grass plants dying after spraying	
In area next to GM crop	**In area 2 km away from GM crop**
83	85

7 (b)   Explain the reason for testing a group of plants that had not been growing near the GM crop.

....

....

                                                            [2]

7 (c)   Suggest **one** way in which the scientists could have made the results of this experiment more reliable.

....

[1]

7 (d) (i) The scientists decided that there was no significant difference between the two groups of plants. Explain whether you agree or disagree with this conclusion.

....

....

[1]

7 (d) (ii) If the scientists are right in their conclusion, does this prove that the concerns about genes for resistance spreading are unfounded? Explain your answer.

....

....

....

....

[2]

7 (e)   Suggest **one** problem that might be caused by the wild grasses becoming resistant to herbicides.

....

....

[1]
[Total 12 marks]

Turn over for the next question

Turn over ▶

8   When a bright light is shone in the eye, the pupil constricts (gets smaller). This is known as the pupillary light reflex and it helps to prevent the eye being damaged by bright light.

A group of scientists decided to investigate the pupillary light reflex in eight volunteers.

The scientists used a computer to measure and record how long it took for the volunteers' pupils to automatically constrict when exposed to a bright light. Each person did the test 20 times and an average reaction time was calculated.

The results are shown in the table.

Name	Average reaction time (s)
Mary	0.29
Tom	0.47
Phillip	0.43
Sarah	0.32
Christopher	0.36
Isobelle	0.23
Lucy	0.41
Ben	0.39

8 (a) (i)   What was the stimulus in this experiment?

....

[1]

8 (a) (ii)   Give **one** way you can tell that the volunteers' response was a reflex.

....

....

[1]

8 (b) Another reflex is the response of moving your hand away from a painful stimulus.

Below is a diagram of the parts of the nervous system involved in this reflex.

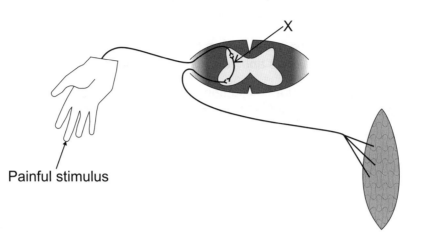

Painful stimulus

8 (b) (i) Name the part labelled **X** on the diagram.

....

[1]

8 (b) (ii) Use words from the box to complete the sentences below about the reflex response.

| stimulus | synapse | receptors | effector | coordinators | muscle | changes |

When the painful stimulus is detected by .... in the skin,

an impulse is sent along a neurone to the central nervous system.

Here, the impulse is passed on to another neurone, via a ....

The impulse is finally transmitted along a motor neurone to the ....

— in this case, a .... which contracts to move the hand away

from the source of pain.

[4]

[Total 7 marks]

Turn over for the next question

9   The chimpanzee and bonobo are two separate species that evolved from a common ancestor.

It is thought that two populations of the ancestor became isolated from each other when the Congo River formed. The population to the north of the river became chimpanzees and the population to the south became bonobos.

9 (a)   *In this question you will be assessed on the quality of your English, the organisation of your ideas and your use of appropriate specialist vocabulary.*

Use your knowledge of natural selection and speciation to suggest how the new species of chimpanzee and bonobo evolved from the same common ancestor.

....

....

....

....

....

....

....

....

....

....

....

....

[6]

9 (b)   In 2005, scientists reported having found fossilised chimpanzee teeth. The teeth were the first known chimpanzee fossils to be found.

Suggest how the teeth became fossilised.

....

....

....

....

[2]

[Total 8 marks]

10   Pigs are often used in agriculture in the UK.

A large proportion of pork produced in the UK comes from pigs that have been reared using intensive farming practices. These practices include keeping the pigs close together in small pens.

10 (a) (i)   The diagram below shows how much of the energy supplied in pig food is transferred to pork meat.

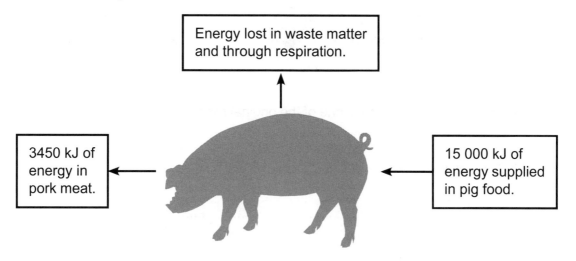

Calculate the percentage of energy in the pig's feed that is lost in waste matter and through respiration.

....

....

....%

[2]

10 (a) (ii)   Suggest how the intensive farming method described above can make the transfer of energy from the pig food to the pork more efficient.

....

....

....

[2]

Question 10 continues on the next page

Turn over ▶

10 (b) Another possible use for pigs is in providing organs suitable for transplant into humans.

The pigs could be genetically engineered to have suitable organs and then cloned using a technique called adult cell cloning.

The cells used in the first stage of this technique are shown in the diagram below.

Cell from Pig **A**
(Unfertilised egg cell)   Cell from Pig **B**
(Adult body cell)

The new pig will be a *clone* of one of the original pig cells.

10 (b) (i) What is a *clone*?

....
[1]

10 (b) (ii) Describe how a pig can be cloned from the cells in the diagram, using adult cell cloning.

....

....

....

....

....

....
[4]

10 (b) (iii) Which of the original pig cells, **A** or **B**, will the new pig be a clone of?

Explain your answer.

....
[1]

[Total 10 marks]

END OF QUESTIONS

Candidate Surname	Candidate Forename(s)

Centre Number	Candidate Number

Level 1/2 Certificate in Biology
Paper 2

Practice Paper
Time allowed: 90 minutes

You must have:
- A ruler.
- A calculator.

Total marks:

Instructions to candidates
- Use **black** ink to write your answers.
- Write your name and other details in the spaces provided above.
- Answer **all** questions in the spaces provided.
- In calculations, show clearly how you worked out your answers.

Information for candidates
- The marks available are given in brackets at the end of each question.
- There are 90 marks available for this paper.
- You should answer Question 5 with continuous prose.
  You will be assessed on the quality of your English,
  the organisation of your ideas and your use of
  appropriate specialist vocabulary.

Advice for candidates
- Read all the questions carefully.
- Write your answers as clearly and neatly as possible.
- Keep in mind how much time you have left.

Get the answers
*Your free Online Edition of this book includes the complete answers and mark scheme for this Exam Paper —
you can even print them out. There's more info about how to get your Online Edition at the front of this book.*

Answer **all** questions

1   Some pondweed was used to investigate how the amount of light available affects the rate of photosynthesis.

The apparatus that was used for this experiment is shown below.

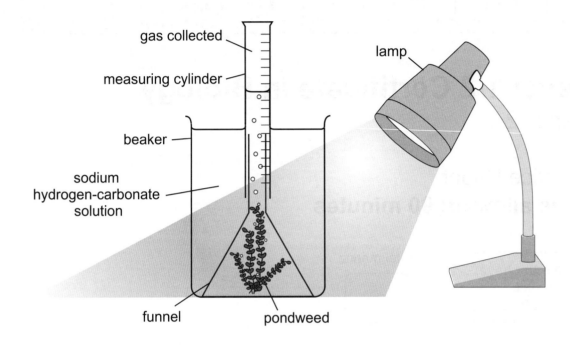

1 (a)   What gas is being collected in the measuring cylinder?

....

[1]

1 (b)   What would happen to the volume of gas collected if the investigation was repeated with the lamp turned off? Give a reason for your answer.

....

....

....

[3]

1 (c)   Sodium hydrogen-carbonate dissolves in water and releases carbon dioxide.
Suggest why sodium hydrogen-carbonate was added to the water in this experiment.

....

....

....

[2]

1 (d) Explain how temperature affects the rate of photosynthesis, and suggest how temperature could be controlled in the experiment.

....

....*Practice...Paper...*

....

....

....

....

[3]

1 (e) The investigation could have been conducted by counting the number of bubbles given off in a certain amount of time by the pondweed.
Suggest **one** advantage of using a measuring cylinder rather than counting bubbles.

....

....

[1]

1 (f) When plants photosynthesise they produce glucose.
Give **three** ways plants use the glucose they produce.

1. ....

2. ....

3. ....

[3]

[Total 13 marks]

Turn over for the next question

Turn over ▶

2   A scientist is investigating how heart rate changes during and after exercise.

The scientist measures the resting heart rate of a volunteer. The volunteer then runs on a treadmill for 2 minutes, then rests for a further 7 minutes. During this time, the scientist measures the volunteer's heart rate every minute with a portable heart rate monitor.

The results are shown in the table below.
The resting heart rate of the volunteer is given at 0 min.

Time after first heart rate measurement taken (min)	Heart rate (beats/min)
0	72
1	118
2	132
3	129
4	116
5	98
6	84
7	76
8	72
9	72

2 (a) (i)   Draw a graph of these results on the graph paper below.
Draw a curve of best fit on your graph.
Add an appropriate scale to the y-axis and label it.

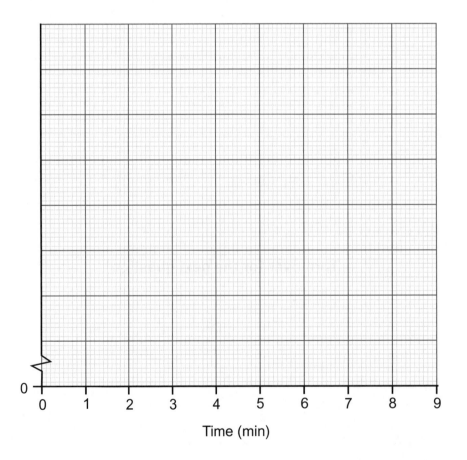

Time (min)

[4]

2 (a) (ii) Use your graph to estimate the volunteer's heart rate after thirty seconds.

Heart rate ....

[1]

2 (a) (iii) The volunteer stopped running two minutes after his first heart rate measurement was taken. How long did it take for his heart rate to return to normal after running?

....

[1]

2 (b) The results of the test suggest that the volunteer's maximum heart rate during the test was 132 beats/min. Suggest **one** way the investigation could be improved to allow the scientist to measure the volunteer's maximum heart rate more accurately.

....

....

[1]

2 (c) When the volunteer first starts running he respires aerobically.

2 (c) (i) Write the word equation for aerobic respiration.

.... + ... → .... + ... (+ energy)

[2]

2 (c) (ii) Give **two** advantages of aerobic respiration over anaerobic respiration.

1. ....

....

2. ....

....

[2]

Question 2 continues on the next page

2 (d) In a second test, the scientist investigates how the concentration of lactic acid in the volunteer's blood changes with different work rates. He produces a graph of his results, as shown below.

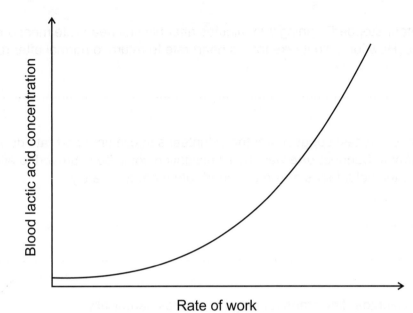

2 (d) (i) Describe the trend shown by the graph.

....

....

....

[2]

2 (d) (ii) Suggest a reason for the trend you described in part **(d) (i)**.

....

....

....

....

....

[3]

2 (d) (iii) After exercise, the volunteer's heart rate remains high for some time.
Explain how this is linked to his blood lactic acid concentration.

....

....

....

....

[2]

[Total 18 marks]

Turn over for the next question

Turn over ▶

3   A student is investigating osmosis.

She sets up an experiment as shown below, putting a different concentration of sucrose solution (measured in moles, M) into each beaker.

She put the same concentration of sucrose solution (0.2 M) into each length of Visking tubing. Visking tubing is a type of artificial, partially permeable membrane.

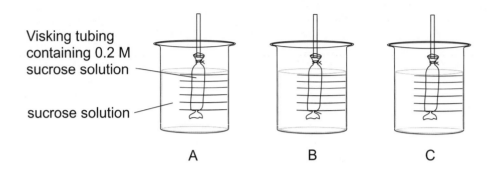

Visking tubing
containing 0.2 M
sucrose solution

sucrose solution

A   B   C

The student records the level of the sucrose solution in each beaker at the start of the experiment and then again after 4 hours. Her results are shown in the table below.

Beaker	Change in level of sucrose solution in beaker (mm)
A	+ 3
B	0
C	− 2

3 (a)   Use information from the table to explain which beaker contained the following concentrations of sucrose solution:

3 (a) (i) 0.2 M

....

....

....

[2]

3 (a) (ii) 0.0 M

....

....

....

[2]

3 (b)   Give **two** variables that need to be kept constant in this experiment.

1. ....

2. ....

[2]

3 (c)   Describe how you would set up a control for this experiment.

....

....

[2]

3 (d)   Suggest **one** way in which the validity of the experiment could be increased.

....

....

[1]

[Total 9 marks]

Turn over for the next question

Turn over ▶

128

4   The global human population is increasing. This has a variety of environmental impacts.

The bar chart below shows the total sulfur dioxide emissions in the UK each year between 1980 and 2000.

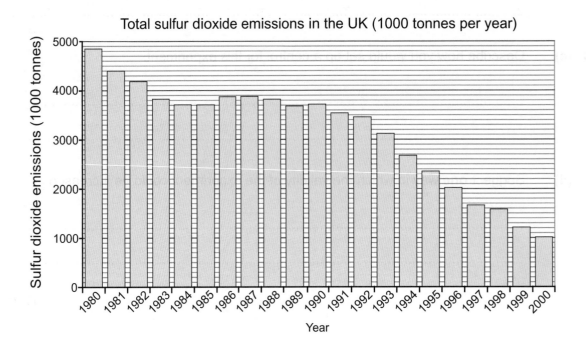

Total sulfur dioxide emissions in the UK (1000 tonnes per year)

Source: National Statistics website: www.statistics.gov.uk.
Crown Copyright material is reproduced with the permission of the Controller of HMSO

4 (a) (i) Describe the trend in the amount of sulfur dioxide emitted between 1980 and 2000.

....

....

....

....

....

[3]

4 (a) (ii) Give **one** environmental problem caused by sulfur dioxide.

....

[1]

4 (a) (iii) Suggest why the level of sulfur dioxide pollution in a country may increase as its population grows.

....

....

[1]

4 (b) The graphs below show the human population size and atmospheric CO_2 concentration between 1500 and 2000.

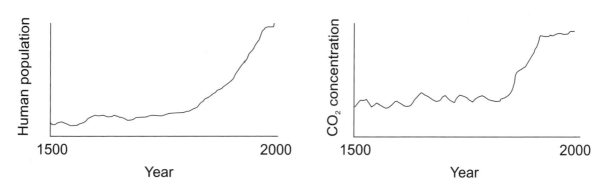

4 (b) (i) Describe the relationship between the two graphs.

....

....

[1]

4 (b) (ii) On their own, these graphs do not prove that the increased human population caused the increased carbon dioxide concentration. Give **three** reasons why not.

1. ....

....

2. ....

....

3. ....

....

[3]

Question 4 continues on the next page

Turn over ▶

4 (c) As the global human population increases, deforestation tends to increase.
Explain how deforestation increases the amount of carbon dioxide in the atmosphere.

....

....

....

[2]

4 (d) The rising concentration of carbon dioxide in our atmosphere has been linked to
global warming. Give **two** possible consequences of global warming.

1. ....

....

2. ....

....

[2]

[Total 13 marks]

5   *In this question you will be assessed on the quality of your English, the organisation of your ideas and your use of appropriate specialist vocabulary.*

The rate at which a plant loses water through its leaves is affected by environmental conditions, such as how windy it is.

The diagram below shows some equipment.

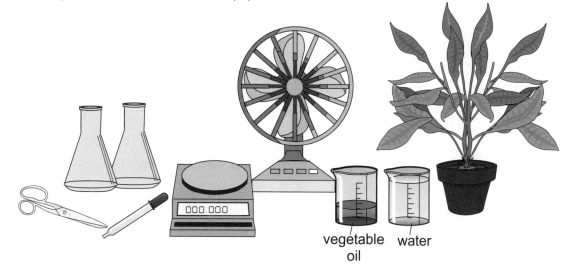

vegetable oil   water

Describe how you could use the equipment shown to investigate the effect of wind (air movement) on the rate of water loss from a plant shoot.

In your answer, you should write about:

• Any measurements you would make.

• The steps you would take to ensure your investigation is valid.

....

....

....

....

....

....

....

....

....

....

....

....

[Total 6 marks]

Turn over for the next question

Turn over ▶

132

6   Quadrats can be used to investigate the distribution of organisms in a habitat.
A group of students used 1 m² quadrats to investigate the plant species that live in
a small field next to their school.

They divided the field up into three sections, as shown in the diagram.

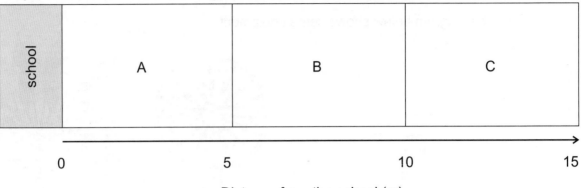

They placed three quadrats at random in each section of the field and counted the
number of buttercups, clover and dandelions in each quadrat.
Their results are shown in the table.

| | Number counted per m² | | | | | | | | | | | |
| | Section A | | | | Section B | | | | Section C | | | |
Plant	1	2	3	Mean	1	2	3	Mean	1	2	3	Mean
Buttercups	50	52	54	52	60	65	64		80	94	90	88
Clover	70	74	69	71	86	96	88	90	121	129	131	127
Dandelions	3	5	4	4	5	7	9	7	13	12	11	12

6 (a)   Calculate the mean number of buttercups per m² in section **B**.

....

....

Mean number of buttercups per m² = ...
[2]

6 (b)   Section **C** measures 5 m by 3 m.
Use the data in the table to estimate the total population of dandelions in section **C**.

....

....

Estimated total population of dandelions in section **C** = ...
[2]

Practice Paper 2

6 (c)   Use the data in the table to draw a conclusion about the distribution of the three plant
species across the field.

....

....

[1]

6 (d)   Give **three** examples of non-living factors that may affect the number of plants of each
species growing at different locations in the field.

1. ....

2. ....

3. ....

[3]

[Total 14 marks]

Turn over for the next question

Turn over ▶

7   In 1970, a new antibiotic was discovered which was very effective against disease X. Doctors have been prescribing this drug ever since. The graph below shows the number of deaths from disease X over a number of years.

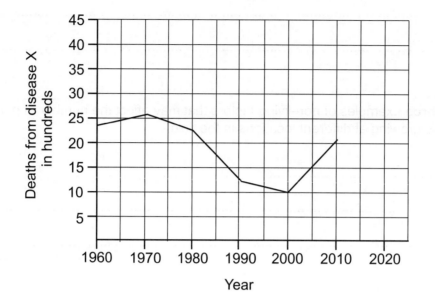

7 (a)   Using the graph, calculate how many more deaths from disease X there were in 1970 than in 2000.

....

.... more deaths.

[1]

7 (b)   Assuming nothing changes, use the graph to predict the number of deaths from disease X in 2020.

Predicted number of deaths in 2020 ....

[1]

7 (c) The diagram below shows two agar plates. Agar plates are Petri dishes containing a culture medium of agar jelly. The bacteria that cause disease X are growing on each plate.

The first plate shows the results of an experiment carried out in **1975**.

The second plate shows the results when the experiment was repeated in **2005** using a fresh sample of bacteria.

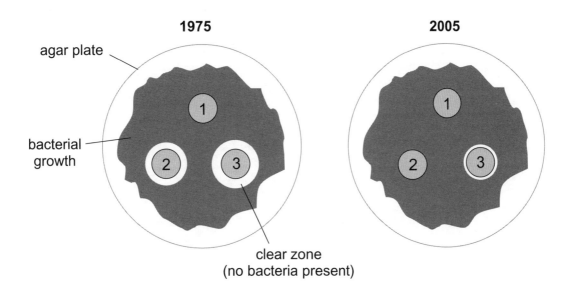

On each plate, there are three paper discs.

- Disc 1 is a control. It is soaked in distilled water only.

- Discs 2 and 3 are soaked in different concentrations of the antibiotic used to treat disease X.

7 (c) (i) Use the evidence from these two plates to suggest reasons for the trends shown in the graph on the previous page.

....

....

....

....

....

....

....

....

[4]

Question 7 continues on the next page

7 (c) (ii) Give **two** steps that scientists carrying out these experiments should have taken in order to prevent the plates being contaminated with unwanted microorganisms.

1. ....

....

2. ....

....

[2]

[Total 8 marks]

8   Many people fear that cod supplies from the North Sea may run out in the near future.
To prevent this, cod stocks need to be fished sustainably.

The graph shows how the population of cod has changed between
1980 and 2005 in the North Sea, just off the coast of Norway.

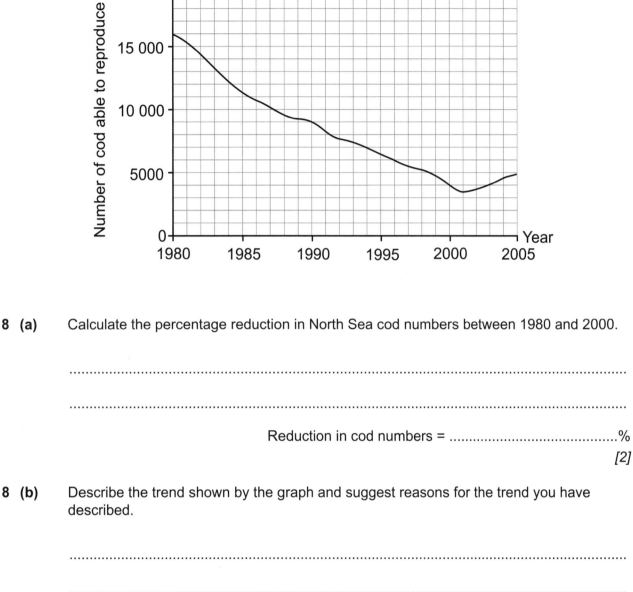

8 (a)   Calculate the percentage reduction in North Sea cod numbers between 1980 and 2000.

....

....

Reduction in cod numbers = ...%

[2]

8 (b)   Describe the trend shown by the graph and suggest reasons for the trend you have
described.

....

....

....

....

....

....

[4]

Question 8 continues on the next page

Turn over ▶

9   A student did an experiment to investigate the effect of temperature on the action of the
     enzyme amylase. The method used is shown below.

> 1.   Add a set quantity of starch solution to a test tube and
>   the same quantity of amylase solution to another.
> 2.   Place the test tubes in a water bath at 10 °C.
> 3.   Allow the starch and amylase solutions to reach the
>   temperature of the water bath, then mix them together
>   and return the mixture to the water bath.
> 4.   Take a small sample of the mixture every minute and
>   test for starch.
> 5.   Stop the experiment when starch is no longer present
>   in the sample, or after 30 minutes (whichever is sooner).
> 6.   Repeat the experiment at different temperatures.

9 (a)   What happens to the starch solution during the experiment?

....

[1]

9 (b)   Name a piece of apparatus that could be used in step 4
         to remove a small sample of the mixture from the test tube.

....

[1]

9 (c)   Explain why a set quantity of starch solution was used for each repeat in the experiment.

....

....

[1]

Question 9 continues on the next page

Turn over ▶

9 (d) The graph below shows the student's results.

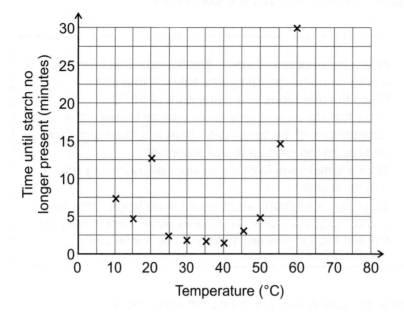

9 (d) (i) Describe and explain the results between 50 °C and 60 °C.

....

....

....

....

....

[3]

9 (d) (ii) The student thinks that one of the results shown on the graph is likely to be anomalous.

Circle the anomalous result on the graph.

Suggest **one** thing the student might have done to cause this anomalous result.

....

....

[2]

[Total 8 marks]

END OF QUESTIONS

Answers

Section One — Cells, Organs and Life Processes

Page 3: Cells

1 a)   E.g. controls the passage of substances into and out of the cell *[1 mark]*
 b)   E.g. where proteins are made *[1 mark]*
 c)   E.g. where most of the (energy-releasing) reactions for respiration take place *[1 mark]*
2 a) i)   X — nucleus *[1 mark]*
         Y — cytoplasm *[1 mark]*
         Z — (permanent) vacuole *[1 mark]*
   ii)   E.g. ———cellulose

[1 mark for an arrow pointing to anywhere on the cell wall]
 b)   Similarity: e.g. both a human cell and a plant cell have a cell membrane/a nucleus/cytoplasm/mitochondria/ribosomes *[1 mark]*.
      Difference: e.g. a human cell doesn't have a permanent vacuole, but a plant cell does. / A human cell doesn't have a cell wall, but a plant cell does. / A human cell doesn't contain chloroplasts, but a plant cell does *[1 mark]*.

Remember most human cells are just like most animal cells.
3   The genetic material is not located in a distinct nucleus in cell B/the bacterial cell *[1 mark]*. There are plasmids present in cell B/the bacterial cell *[1 mark]*.

Page 4: Specialised Cells

1 a) i)   E.g. it has a long tail/streamlined head *[1 mark]* to help it to swim to the egg *[1 mark]*.
   ii)   To provide the energy needed to swim to the egg *[1 mark]*.
 b)   To nourish/provide food for the developing embryo *[1 mark]*.
 c)   E.g. it contains the DNA/genetic material that will go into the developing embryo *[1 mark]*. / It controls the cell's activities *[1 mark]*.
2   E.g. you would expect white blood cells to have lots of ribosomes *[1 mark]* to allow them to produce large quantities of antibodies/proteins *[1 mark]*.

Pages 5-6: Tissues, Organs and Organ Systems

1 a) i)   Name: liver *[1 mark]*
         Function: e.g. produces bile *[1 mark]*
   ii)   Name: large intestine *[1 mark]*
         Function: absorbs water from undigested food, leaving faeces *[1 mark]*.
   iii) Name: small intestine *[1 mark]*
         Function: absorbs soluble food molecules *[1 mark]*.
 b) i)   Pancreas *[1 mark]* and salivary glands *[1 mark]*.
   ii)   E.g. muscular tissue, which contracts and moves food through the digestive system / epithelial tissue, which lines the interior and exterior of the stomach.
         [1 mark for the type of tissue, 1 further mark for a correct description of its function]
2 a)   A group of organs working together to perform a particular function *[1 mark]*.
 b)   E.g. glandular tissue *[1 mark]* because this sort of tissue secretes hormones *[1 mark]*.
 c) i)   Muscular tissues contracts *[1 mark]*, which may help to push/transport urine down into the bladder *[1 mark]*.
   ii)   Because they both contain different tissues working together to perform a particular function *[1 mark]*.
 d)   In order to become specialised to carry out particular functions *[1 mark]*.

Page 7: Plant Tissues and Organs

1 a)   E.g. stems *[1 mark]*, roots *[1 mark]*, leaves *[1 mark]*
 b) i)   epidermal tissue *[1 mark]*
   ii)   Spongy mesophyll *[1 mark]*. This type of tissue contains air spaces to allow gases to diffuse in and out of cells *[1 mark]*. Palisade mesophyll *[1 mark]*. This is the part of the leaf where most photosynthesis happens *[1 mark]*.
2 a) i)   A: phloem *[1 mark]*
         B: xylem *[1 mark]*
   ii)   E.g. to transport water, mineral ions and sucrose around the plant *[1 mark]*.
 b)   E.g. it might have root hairs around the outside *[1 mark]*. / The xylem and phloem tissue would both be in the centre of the root *[1 mark]*. / The xylem and phloem would not be arranged in bundles around the edge of the root *[1 mark]*.

Page 8: Diffusion

1 a)   high(er), low(er) *[1 mark]*
 b)   It will cause it to decrease *[1 mark]*.
 c)   respiration *[1 mark]*
2 a)

[1 mark for dye particles spread out in the water]
 b)   The rate of diffusion would be faster *[1 mark]* because the larger the difference in concentration is, the faster the rate of diffusion *[1 mark]*.
3   Z particles must be larger than X and Y particles *[1 mark]* because larger particles will not diffuse through the membrane *[1 mark]*.

Page 9: Osmosis

1 a) i)   E.g. osmosis is the movement of water molecules across a partially permeable membrane *[1 mark]* from a region of high water concentration to a region of low water concentration *[1 mark]*.
   ii)   The net movement of water will be into the body cells *[1 mark]* because the tissue fluid has a higher water concentration than the body cells *[1 mark]*.
 b)   A hypotonic solution *[1 mark]* because the tissue fluid has a higher concentration of water than the fluid inside the cells *[1 mark]*.
 c)   The net movement of water molecules stops when there is an equal concentration of water molecules/an isotonic solution on either side of the membrane *[1 mark]*.

There needs to be a concentration gradient (i.e. a difference in the concentration of water molecules on either side of a membrane) for osmosis to take place.
2 a)   A — normal
      B — turgid
      C — flaccid
      D — plasmolysed
      [2 marks if all 4 answers correct, 1 mark if at least 2 answers are correct]
 b)   The plant cells lose water and their contents are no longer able to push up against the inelastic cell wall *[1 mark]*. Turgor pressure is lost, so the plant goes limp/wilts *[1 mark]*.

Pages 10-11: Diffusion and Osmosis Experiments

1 a)   diffusion *[1 mark]*
 b) $(835 + 825 + 842 + 838) \div 4 =$ **835 s**
      [2 marks for correct answer, otherwise 1 mark for correct working]

c) As the size of the gelatine cube increases, the time taken for the cube to become yellow increases *[1 mark]*.

2 a) E.g. they could measure the mass of each egg before putting it in its jar and measure each egg's mass again after one day. *[1 mark for stating what will be measured, 1 mark for stating a time period over which it will be measured.]*

b) E.g. the egg in the weak sugar solution would lose mass and the egg in water would gain mass by the end of the experiment *[1 mark for describing a result in the weak sugar solution which suggests water has moved out of the egg, 1 mark for describing a result in the water which suggests water has moved into the egg.]*

c) E.g. a third egg, weighed then placed back in the jar of vinegar *[1 mark]*.

3 a) $(0.78 + 0.81 + 0.82) \div 3 = \textbf{0.80 g}$
[2 marks for correct answer, otherwise 1 mark for correct working]

b) 0.4 M *[1 mark]*. The mean change in mass at this concentration is 0, which suggests that the concentration of the sugar solution and the concentration of the fluid inside the potato are the same (or very similar) *[1 mark]* because water hasn't entered the potato cylinders via osmosis *[1 mark]*.

c) To show that results are repeatable / to improve validity of the results *[1 mark]*.

d) E.g. the mass of potato cylinders *[1 mark]* / the shape/surface area of the potato cylinders *[1 mark]* / the volume of sugar solution used *[1 mark]* / the temperature each beaker is kept at *[1 mark]*.

Page 12: Active Transport

1 a) root hair cell *[1 mark]*

b) The soil generally has a lower concentration of minerals than the cells in the root. / The cells in the root contain a higher concentration of minerals than the soil *[1 mark]*. Diffusion only takes place from areas of high concentration to areas of low concentration *[1 mark]*, so active transport is needed to move minerals ions against the concentration gradient into the root hair cells *[1 mark]*.

c) i) Active transport requires energy from respiration *[1 mark]* and most of the respiration reactions take place in the mitochondria *[1 mark]*.

ii) E.g. it has a large surface area *[1 mark]*.
The long protruding 'hair' increases the cell's surface area.

2 a) E.g. to absorb nutrients/glucose into the bloodstream *[1 mark]* when there is a low concentration of nutrients/glucose in the intestine and a higher concentration of nutrients/glucose in the blood *[1 mark]*.

b) E.g. to absorb glucose/sugar *[1 mark]* from low concentrations in the kidney tubules *[1 mark]*.

Page 13: Gas and Solute Exchange

1 a) large, small *[1 mark]*

b) In single-celled organisms like *Euglena*, substances can diffuse straight into and out of the cell (across the cell membrane) *[1 mark]*. The rate of diffusion is quick because of the short distances substances have to travel/the relatively large surface area to volume ratio of single-celled organisms *[1 mark]*. But in multicellular organisms like trout, diffusion across the outer surface would be too slow to reach cells deep in the organism's body/diffusion across the outer surface would be too slow because of the organism's relatively small surface area to volume ratio *[1 mark]*. So multicellular organisms need specialised exchange organs to move substances to and from individual cells quickly *[1 mark]*.

c) Any two from: e.g. the gill filaments have a large surface area *[1 mark]* so large quantities of gases can diffuse at once *[1 mark]*. / The gill filaments/lamallae are thin *[1 mark]* so gases only have a short distance to diffuse *[1 mark]*. / The gills have an efficient blood supply *[1 mark]* so gases can enter and leave the blood quickly *[1 mark]*.

Page 14: Exchanging Materials in Mammals

1 a) Any three from: e.g. they have thin outer walls *[1 mark]*. / They have a good blood supply *[1 mark]*. / They have a moist lining *[1 mark]*. / They have a very large surface area *[1 mark]*. / They are ventilated *[1 mark]*.

b) E.g. carbon dioxide diffuses out of the blood in the capillaries and into the alveoli *[1 mark]*. Oxygen diffuses into the blood from the air inside the alveoli *[1 mark]*.

2 a) small intestine *[1 mark]*

b) i) E.g. a single, permeable layer of surface cells *[1 mark]*. A good blood supply/network of capillaries *[1 mark]*.

ii) Having villi means that there is a much larger surface area for the absorption of nutrients *[1 mark]*.

c) If the villi become flattened, this will reduce their surface area for the absorption of nutrients *[1 mark]*, which could mean that not enough vitamins are absorbed, leading to low vitamin levels in the body *[1 mark]*.

Page 15: Genes and Chromosomes

1 1 gene, 2 chromosome, 3 nucleus, 4 cell
[2 marks if all 4 answers correct, 1 mark if at least 2 answers are correct]

2 nucleus *[1 mark]*, genes *[1 mark]*, pairs *[1 mark]*

3 a) E.g. a short section of DNA that controls the development of a particular characteristic *[1 mark]*.

b) There are different versions of the same gene, called alleles *[1 mark]*, that give different versions of a characteristic *[1 mark]*. The two kittens must have different versions/ alleles of the gene for fur length, meaning one is long-haired and the other is short-haired *[1 mark]*.

Page 16: Cell Division — Mitosis

1 a) once *[1 mark]*, two *[1 mark]*, identical *[1 mark]*, copies *[1 mark]*

b) To produce new cells for growth *[1 mark]*. To replace cells that have been damaged *[1 mark]*.

2 a) The amount of DNA is doubling *[1 mark]* because the DNA is being copied/duplicating itself *[1 mark]*.

b) The new cells separate *[1 mark]*.

Page 17: Cell Division — Meiosis

1 a) i) three *[1 mark]*
ii) two *[1 mark]*
iii) four *[1 mark]*
When a cell undergoes meiosis, it divides twice to form four gametes, each with half the normal number (a single set) of chromosomes — it doesn't matter whether you're talking about human cells or mosquito cells.

b) the reproductive organs *[1 mark]*

c) Before a cell divides by meiosis, the cell copies/duplicates its DNA *[1 mark]*.

2 a) So that when two gametes join together during fertilisation the resulting fertilised egg will have the full set of chromosomes *[1 mark]*.

b) The fertilised egg cell divides repeatedly by mitosis to form many new cells *[1 mark]*. These cells differentiate during development to form different kinds of specialised cells *[1 mark]*.

Page 18: Stem Cells and Differentiation

1   lose *[1 mark]*, plant *[1 mark]*, stem cells *[1 mark]*

2 a)   E.g. paralysis *[1 mark]* could potentially be treated using stem cells that are made to differentiate into nerve cells *[1 mark]*.

   b)   How to grade your answer:

       0 marks: No relevant information is given.

       1-2 marks: A comparison is made between the potential use of the two types of stem cell.

       3-4 marks: Two clear comparisons are made between the potential use of the two types of stem cell. The answer has a logical structure and spelling, grammar and punctuation are mostly correct.

       5-6 marks: At least three clear, detailed comparisons are made between the potential use of the two types of stem cell. The answer has a logical structure and uses correct spelling, grammar and punctuation.

       Here are some points your answer may include:

       Adult stem cells are able to differentiate into a smaller range of cell types than embryonic stem cells, which are capable of developing into any type of cell. This means embryonic stem cells may be used to treat a wider range of conditions than adult stem cells.

       Adult stem cells are relatively rare compared with embryonic stem cells. Adult stem cells are also harder to grow in culture then embryonic stem cells, so it may be easier to produce large numbers of embryonic stem cells.

       Adult stem cells (which come from the patient's own body) are less likely to be rejected than embryonic stem cells, so the patient receiving them doesn't need to take medication to stop rejection occurring. However therapeutic cloning offers a way to produce embryonic stem cells that won't be rejected by the body.

       There are ethical issues with using embryonic stem cells, which don't apply to using adult stem cells. For example, some people feel that human embryos shouldn't be used as a source of stem cells as each one is a potential human life.

Page 19: Growth of Tumours

1 a)   To repair and replace damaged tissue *[1 mark]*.

   b)   In benign tumours, the cancerous cells do not spread to other sites in the body/invade healthy tissues, but in malignant tumours they can do *[1 mark]*.

   c)   Malignant cancer cells/cells from malignant tumours *[1 mark]* enter the bloodstream and circulate to other parts of the body, where they form secondary tumours *[1 mark]*.

   d) i)   E.g. tobacco smoke *[1 mark]*, asbestos *[1 mark]*

A chemical carcinogen is a chemical that causes cancer.

     ii)   E.g. UV rays/X-rays *[1 mark]*

2 a)   E.g. the results show that the relative risk of breast cancer for women in California is higher for women who previously smoked or still smoke compared to those who have never smoked *[1 mark]*. The relative risk is highest overall for those who still smoke *[1 mark]*.

For this question it's important that you make a comparison between the results for women who have never smoked and those who have or still do — this is what the study was aiming to investigate.

   b)   That for women in California, smoking is associated with an increased risk of breast cancer *[1 mark]*.

Section Two — Breathing and Circulation

Pages 20-21: The Respiratory System

1 a)   To get air into and out of the lungs *[1 mark]* so that oxygen can diffuse out of the air into the bloodstream *[1 mark]* and carbon dioxide can diffuse out of the bloodstream into the air *[1 mark]*.

   b) i)   **A** = bronchus, **B** = trachea, **C** = intercostal muscle *[1 mark for each correct answer]*

     ii)   diaphragm *[1 mark]*

   c)   The intercostal muscles contract, pulling the ribcage up *[1 mark]*. The diaphragm contracts and flattens out *[1 mark]*. This increases the thorax volume, which decreases the pressure inside the lungs so it's less than the air pressure outside *[1 mark]*. This draws air into the lungs *[1 mark]*.

2   The alveoli are the site of gas exchange in the lungs *[1 mark]*. Oxygen diffuses from the air into the bloodstream via the capillaries surrounding the alveoli *[1 mark]*. Destruction of these capillaries and a reduction in the surface area of the alveoli, mean that less oxygen is able to diffuse into the bloodstream from the air *[1 mark]*.

3 a) i)   A positive pressure ventilator works by pumping air into the lungs, which expands the ribcage *[1 mark]*. When it stops pumping, the ribcage relaxes and pushes air back out of the lungs *[1 mark]*.

     ii)   E.g. the woman may be unable to breathe by herself *[1 mark]* because of a (lung) injury/disease *[1 mark]*.

   b) i)   The drop in pressure makes the lungs expand/the thorax volume increase, drawing air into the lungs *[1 mark]*.

     ii)   Air is pumped back into the case to increase the pressure around the chest *[1 mark]*. This decreases the volume of the thorax, forcing air out of the lungs *[1 mark]*.

It makes sense that if air is pumped out of the case to decrease the pressure around the chest and draw air into the lungs, the opposite will happen to make air leave the lungs.

Page 22: Respiration

1 a)   To release energy *[1 mark]*.

   b) i) $C_6H_{12}O_6 + 6O_2 \rightarrow 6CO_2 + 6H_2O$ (+ energy)
[1 mark for $C_6H_{12}O_6 + 6O_2$ on the left-hand side of the equation, 1 mark for $6CO_2 + 6H_2O$ on the right. Allow 1 mark if the correct symbols are used, but the equation isn't correctly balanced.]

     ii)   It provides a continuous supply of energy for the organism's cells *[1 mark]*.

   c)   Plants use the energy from aerobic respiration to build sugars, nitrates and other nutrients *[1 mark]* into amino acids, which are then built into proteins *[1 mark]*.

2   A rat needs energy from (aerobic) respiration to maintain its body temperature in colder surroundings *[1 mark]*. (Aerobic) respiration uses glucose, which the rat obtains from its food *[1 mark]*. A snake cannot maintain its body temperature in colder surroundings, so it may need less energy and therefore less food than a rat *[1 mark]*.

3   There are more mitochondria in the muscle cells than in the skin cell *[1 mark]*. Mitochondria are the site of most of the reactions of aerobic respiration, which releases energy for the cell *[1 mark]*. The muscle cells need more mitochondria than the skin cell because muscle cells need a lot of energy to contract *[1 mark]*.

Pages 23-24: Respiration and Exercise

1 a)   E.g. during vigorous exercise *[1 mark]* because the body is unable to supply enough oxygen to the muscles *[1 mark]*.

   b) $C_6H_{12}O_6 \rightarrow 2C_3H_6O_3$ (+ energy)
[1 mark for $C_6H_{12}O_6$ on the left-hand side of the equation, 1 mark for $2C_3H_6O_3$ on the right-hand side of the equation. Allow 1 mark if the correct symbols are used, but the equation isn't correctly balanced.]

   c)   E.g. it releases much less energy (per glucose molecule) than aerobic respiration *[1 mark]*. Lactic acid builds up in the muscles causing muscle fatigue *[1 mark]*.

   d)   ethanol *[1 mark]*, carbon dioxide/CO_2 *[1 mark]*

2 a)   During strenuous exercise, the muscle cells start respiring anaerobically *[1 mark]*. This produces lactic acid, increasing the blood lactate level *[1 mark]*.

   b)   The blood lactate level decreases as the lactic acid is being oxidised *[1 mark]* to carbon dioxide and water *[1 mark]*.

3 a) $45 - 15 = \mathbf{30}$ *[1 mark]*
b)   During exercise the runner's muscles respire more/require more energy *[1 mark]* so her breathing rate increases to supply more oxygen to her cells *[1 mark]* and remove more carbon dioxide from her body *[1 mark]*.
c)   Because she has an oxygen debt after exercising *[1 mark]*. Extra oxygen is needed to break down the lactic acid produced by anaerobic respiration *[1 mark]*, so she needs to keep breathing hard to get more oxygen into her blood *[1 mark]*.
d)   E.g. her heart rate will increase *[1 mark]*, to increase blood flow to her muscles *[1 mark]*.
4   E.g. during vigorous exercise, the muscles use glucose rapidly *[1 mark]*. Blood glucagon level may rise so that some of the glycogen stored in the liver can be converted back to glucose to provide more energy *[1 mark]*.

Pages 25-26: Circulatory System — The Heart

1 a)   A: aorta *[1 mark]*
B: vena cava *[1 mark]*
C: left atrium *[1 mark]*
b) i)   the right atrium *[1 mark]*.
   ii)   E.g. an artificial pacemaker could be inserted under the skin *[1 mark]*.
c)   To prevent blood flowing backwards from the right ventricle to the right atrium *[1 mark]*.
d)   muscle *[1 mark]*
e)   Deoxygenated blood arrives at the heart through the vena cava *[1 mark]* and enters the (right) atrium *[1 mark]*. The blood is then pumped into the (right) ventricle *[1 mark]*, which contracts to pump the blood to the lungs through the pulmonary artery *[1 mark]*.
2 a)   Arteries, because blood is being transported away from the heart *[1 mark]*.
b)   To transport nutrients and oxygen to every cell in the body *[1 mark]* and to carry waste products (such as carbon dioxide and urea) to where they can be removed from the body *[1 mark]*.
c) i)   Any two from: e.g. a fish has a single circulatory system, whereas a human has a double circulatory system/in a fish, blood is transported around the body in a single circuit, whereas in a human there are two circuits (one to the lungs and one to the rest of the body) *[1 mark]*. / Blood enters the heart at a single place in a fish and at two places in a human *[1 mark]*. / Blood leaves the heart at a single place in a fish and at two places in a human *[1 mark]*. / Blood is pumped to the gills in a fish and to the lungs in a human *[1 mark]*.
   ii)   E.g. both circulatory systems involve a heart/blood *[1 mark]*. / Both circulatory systems pump blood around the body *[1 mark]*.

Page 27: Circulatory System — Blood Vessels

1 a)   To carry the blood close to the body cells *[1 mark]*, to exchange substances such as oxygen and carbon dioxide *[1 mark]*.
b)   Any two from: e.g. a capillary is very narrow *[1 mark]* so it can carry blood close to any cell *[1 mark]*. / It has permeable walls *[1 mark]* so substances can diffuse in and out of them *[1 mark]* / Its walls are thin/only one cell thick *[1 mark]* to increase the rate of diffusion by decreasing the distance over which it happens *[1 mark]*.
2 a) i)   To prevent the backflow of blood. / To help keep blood flowing in the right direction *[1 mark]*.
   ii)   E.g. veins have a bigger lumen/thinner wall *[1 mark]*. / Arteries has a smaller lumen/thicker wall *[1 mark]*.

To answer this question, you needed to think about how the structure of veins and arteries differ — and how you could tell them apart by just looking at them.

b)   The vein, because it has a thinner wall *[1 mark]*.
c)   The length of the blood vessel *[1 mark]* because this is the dependent variable/the variable that is measured (which is always plotted on the vertical axis) *[1 mark]*.

Page 28: Circulatory System — The Blood

1 a)   They help to clot the blood at the site of a wound *[1 mark]*.
b)   red blood cells *[1 mark]*, white blood cells *[1 mark]*
c)   Any two from: e.g. soluble products of digestion/nutrients from the gut/small intestine to the body cells / carbon dioxide from the organs to the lungs / urea from the liver to the kidneys *[1 mark for each correct answer]*.
2 a)   E.g. this cell does not have a nucleus, whereas a cell involved in defending the body against disease (a white blood cell) does *[1 mark]*.
b)   haemoglobin, oxyhaemoglobin, e.g. tissues/other organs *[1 mark for each correct answer]*
3   In the normal blood clotting process, fibrinogen is converted into fibrin *[1 mark]*, which forms a mesh to trap platelets and (red) blood cells, forming a blood clot *[1 mark]*. If fibrinogen is absent or only present in low levels, the blood may not clot properly, leading to excessive bleeding from a wound *[1 mark]*.

Page 29: Blood Type and Organ Donation

1 a)   A protein on the surface of a cell *[1 mark]*.
b)   Because the antigens are foreign/not recognised by the patient's immune system *[1 mark]*.
c)   E.g. a donor with a tissue type that closely matches the patient should be chosen *[1 mark]*. The patient should take drugs to suppress their immune system *[1 mark]*.
2 a)

Blood Group	Can give blood to:	Can get blood from:
A	A and AB	A and O
B	B and AB	B and O
AB	only AB	anyone
O	anyone	only O

[1 mark for each correct answer]

b)   Group A blood contains anti-B antibodies *[1 mark]*. If it came into contact with group B antigens it would cause agglutination/blood clumping *[1 mark]*.
c)   There are no antigens on group O red blood cells *[1 mark]*, so the cells can't be attacked by anti-A or anti-B antibodies/ antibodies from A, B or AB blood groups *[1 mark]*.

Page 30: Circulation Aids

1   coronary arteries, fatty, muscle, (e.g.) oxygen *[1 mark for each correct answer]*
2 a)   E.g. damage may cause the heart valve to stiffen, so it won't open properly *[1 mark]*. / Damage may cause the heart valve to leak, allowing blood to flow in both directions *[1 mark]*.
b)   E.g. it could be replaced with a biological heart valve / a valve taken from another human/mammal *[1 mark]*.
3   How to grade your answer:
0 marks: No relevant information is given.
1-2 marks: Brief descriptions of at least one advantage and one disadvantage are given.
3-4 marks: At least two advantages and two disadvantages are given. The answer has a logical structure and spelling, grammar and punctuation are mostly correct.

5-6 marks: The answer gives at least three advantages and three disadvantages. The answer has a logical structure and uses correct spelling, grammar and punctuation.

Here are some points your answer may include:

Advantages:

An artificial heart is less likely to be rejected by the body's immune system than a donor heart.

A patient receiving an artificial heart does not need to take drugs to suppress their immune system, unlike with a donor heart.

There is no need to wait for an artificial heart to become available, unlike with a donor heart.

Disadvantages:

Artificial hearts do not tend to work as well as healthy donor hearts — parts of a mechanical heart can wear out or the electric motor could fail.

Blood does not tend to flow as smoothly through an artificial heart as it does with a donor heart, which can cause blood clots/strokes.

A patient with an artificial heart has to take drugs to thin their blood to prevent clots, unlike with a donor heart. This can leading to bleeding problems if the patient is hurt in an accident.

The question asks you to discuss the advantages and disadvantages of an artificial heart over a donor heart — so make sure you give a balanced account of each. Also make sure you make clear comparisons between the two treatments, e.g. 'There is no need to wait for an artificial heart to become available, <u>unlike with a donor heart.</u>'

Section Three — Enzymes and Digestion

Page 31: Carbohydrates, Lipids and Proteins

1   glycerol *[1 mark]*, fatty acids *[1 mark]*
2 a)   simple *[1 mark]*, e.g. Glucose/Sucrose *[1 mark]*
  b) i)   e.g. cellulose *[1 mark]*
     ii) A long chain of simple sugar units joined together *[1 mark]*.
3 a)   Proteins/antibodies are made of long chains of amino acids *[1 mark]*. These fold into a specific shape to enable a specific molecule/antigen to fit into the protein/antibody *[1 mark]*.
  b)   e.g. enzymes *[1 mark]*, hormones *[1 mark]*, structural components (of tissues such as muscle) *[1 mark]*

Pages 32-33: Enzymes

1 a)

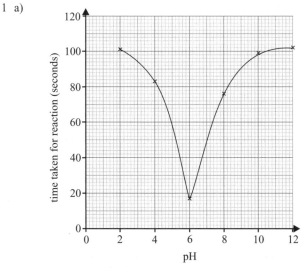

[1 mark for labelling the y axis correctly and using a suitable scale, 1 mark for using straight lines to join the points, 2 marks for correctly plotted points to within ± 1 second. Deduct up to 2 marks for incorrectly plotted points, 1 mark per incorrect point.]

  b)   pH 6 *[1 mark]*
The optimum pH is the pH at which the reaction happens fastest.
  c)   If the bonds are broken, the shape of the active site may change/the enzyme may denature *[1 mark]*, meaning that it can't speed up the reaction *[1 mark]*.
  d)   Any two from: e.g. the temperature should be the same at each pH / the same volume of the reactant and enzyme should be used for each pH / the same method of determining when the reaction is complete should be used for each pH / everything should be measured and timed as accurately as possible using appropriate equipment *[1 mark for each correct answer, maximum 2 marks]*.
  e)   The reaction would be much slower *[1 mark]* because the enzyme is a catalyst that speeds up the chemical reaction *[1 mark]*.
2 a)   35 °C (accept ± 1 °C) *[1 mark]*
  b)   At 45 °C the enzyme's active site has changed shape/the enzyme has denatured *[1 mark]*. This means the substrate no longer fits into the active site and so the reaction stops *[1 mark]*.
3   0 marks: There is no relevant information.
      1-2 marks: There is a brief description of a method with the steps in approximately the correct order.
      3-4 marks: The method is described clearly and would give valid results. The answer has a logical structure and spelling, punctuation and grammar are mostly correct.
      5-6 marks: The method is described clearly and in detail, so that it would give valid results. The answer has a logical structure and uses correct spelling, grammar and punctuation.

Here are some points your answer may include:

Add a set volume of hydrogen peroxide solution to a test tube.

Put the test tube in a water bath so that the temperature can be controlled.

Add five potato discs of the same thickness and diameter to the test tube.

Quickly attach a bung to the top of the test tube and connect a gas syringe.

Use the gas syringe to collect and measure the volume of oxygen given off over a set period of time, measured with a stop watch.

Repeat the experiment three times at a minimum of three different temperatures, using discs cut from the same potato each time.

Make sure that the same concentration and volume of hydrogen peroxide solution are used each time.

Make sure that the number of potato discs is used each time.

Make sure that the same thickness and diameter of potato disc is used each time.

Use the average volume of oxygen collected to calculate the rate of catalase activity in potato cells at each temperature.

In the exam you might well be asked to describe how to do an investigation that you've not come across before. If you do, don't panic — make sure you really understand what your investigation needs to find out and look at the apparatus and materials available for clues about the sort of experiment you could do.

Pages 34-35: Enzymes and Digestion

1 a)   proteins *[1 mark]*, amino acids *[1 mark]*, small intestine *[1 mark]*
  b)   Amylase breaks down starch into sugars *[1 mark]*.
  c)

Location	Protease	Lipase	Amylase
Salivary glands			✓
Stomach	✓		
Small intestine	✓	✓	✓

[1 mark for each correct row.]

..

2 a) If the pancreas isn't functioning properly it may not produce enough lipase *[1 mark]* to break down all the lipids/fats in the diet, so lots may be excreted in the faeces *[1 mark]*.

b) E.g. it may be more difficult to digest starch *[1 mark]* because the pancreas may produce less amylase *[1 mark]*. / It may be more difficult to digest proteins *[1 mark]* because the pancreas may produce less protease *[1 mark]*.

3 Molecules such as starch, proteins and lipids can't pass through the walls of the digestive system because they are large and insoluble *[1 mark]*. Digestive enzymes break them down into smaller, soluble molecules *[1 mark]* that can be absorbed into the blood through the wall of the small intestine *[1 mark]*.

4 a) i) bile *[1 mark]*
   ii) Produced: liver *[1 mark]*
       Acts: small intestine *[1 mark]*
   iii) The enzymes in the small intestine (where the bile/fluid acts) work best in alkaline conditions *[1 mark]*. The alkaline bile/fluid neutralises the acid from the stomach, so the enzymes can work *[1 mark]*.

b) Bile emulsifies fat/breaks fat down into tiny droplets *[1 mark]*. This gives a larger surface area for lipases to work on and so the fat is digested more quickly *[1 mark]*. The gallstones could block the bile ducts and prevent bile from entering the small intestine *[1 mark]*. If so, any fat may be digested more slowly, possibly causing problems *[1 mark]*.

Page 36: More on Enzymes and Digestion

1 Exoenzymes, because they are produced by specialised cells in glands and in the gut lining *[1 mark]* and are released into the gut to mix with food *[1 mark]*.

2 a) A — salivary glands *[1 mark]*
     B — gall bladder *[1 mark]*
     C — duodenum *[1 mark]*
     D — pancreas *[1 mark]*

b) i) hydrochloric acid *[1 mark]*
   ii) Acidic conditions give the right pH for protease/digestive enzymes in the stomach to work *[1 mark]*.

c) To absorb most of the water that is mixed with food into the bloodstream *[1 mark]*.

Page 37: Uses of Enzymes

1 a) i) Protease/Protein-digesting enzymes can be used to pre-digest proteins in baby foods *[1 mark]*. This makes them easier for the baby to digest *[1 mark]*.

   ii) An isomerase enzyme can be used to convert glucose syrup into fructose syrup *[1 mark]*. Fructose is sweeter than glucose *[1 mark]*. This means less fructose is needed to achieve the same level of sweetness in the slimming foods, so the food contains fewer calories *[1 mark]*.

b) Carbohydrases are used to convert starch syrup to sugar syrup *[1 mark]*.

2 a) They produce the enzymes that are used in the washing powders *[1 mark]*.

b) A, because it contains lipase enzymes which digest fats/lipids *[1 mark]*.

c) The enzymes in a biological washing powder mean that it works better at low temperatures than a non-biological washing powder *[1 mark]*. Washing at lower temperatures uses less energy which reduces the energy costs of doing laundry *[1 mark]*.

Section Four — Nerves and Homeostasis

Page 38: The Nervous System

1 a) Stimuli are changes in the environment (that you may need to react to) *[1 mark]*.
  b) receptors *[1 mark]*
  c) i) the eyes *[1 mark]*
     ii) the skin *[1 mark]*
     ii) the ears *[1 mark]*

2 a) (sensory) neurone *[1 mark]*
  b) brain *[1 mark]*, spinal cord *[1 mark]*
  c) E.g. a muscle *[1 mark]* responds by contracting *[1 mark]*. A gland *[1 mark]* responds by secreting chemicals *[1 mark]*.

3 Stimulus: the smell of the food/chemicals in the air *[1 mark]*
  Sense organ: the nose *[1 mark]*
  Effectors: the salivary glands *[1 mark]*

Page 39: Reflexes

1 a) X: sensory neurone *[1 mark]*
     Y: relay neurone *[1 mark]*
     Z: motor neurone *[1 mark]*
  b) the spinal cord *[1 mark]*
  c) i) a synapse *[1 mark]*
     ii) E.g. because the nervous impulse is transferred by chemicals, which diffuse/move across the gap *[1 mark]* and this is a relatively slow process/takes more time than the transmission of an impulse along a nerve *[1 mark]*.
  d) The effector is the muscle *[1 mark]* and it responds by contracting (which causes the man to drop the plate) *[1 mark]*.
  e) Because a reflex is a fast *[1 mark]*, automatic reaction (so he didn't need to think about what he was doing) *[1 mark]*.

Page 40: The Brain

1 a) The medulla *[1 mark]* because this part of the brain controls unconscious activities, such as breathing *[1 mark]*.
  b) i) X *[1 mark]*
     ii) Any two from: e.g. consciousness *[1 mark]* / intelligence *[1 mark]* / memory *[1 mark]* / language *[1 mark]*

2 a) E.g. the cerebellum *[1 mark]* as this is the region responsible for the muscle coordination *[1 mark]*.
  b) E.g. studying patients with brain damage *[1 mark]*, electrically stimulating different regions of the brain *[1 mark]*, MRI scans *[1 mark]*.

Page 41: Homeostasis

1 a) i) $1.2 \times 2 = 2.4$
        $2.4 \div 0.8 = 3$ bottles *[2 marks for the correct answer, otherwise 1 mark for the correct working]*
     ii) Via the lungs in his breath *[1 mark]*.
  b) In the blood *[1 mark]*.
  c) The cyclist will also be losing ions in his sweat *[1 mark]* and using up glucose, both of which can be replaced by the sports drink but not by water *[1 mark]*.

2 a) (the thermoregulatory centre in) the brain *[1 mark]*
  b) the woman's skin *[1 mark]* and (the thermoregulatory centre in) her brain *[1 mark]*
  c) So that the effectors can respond to increase her body temperature back to a normal level *[1 mark]*.

Page 42: The Liver and Kidneys

1 a) The liver breaks down excess amino acids into ammonia *[1 mark]* by a process called deamination *[1 mark]*. This ammonia is then converted into urea *[1 mark]*.
  b) i) detoxification *[1 mark]*
     ii) They are removed in the urine by the kidneys *[1 mark]*
  c) iron *[1 mark]*

2 a) Any two from: e.g. water *[1 mark]* / urea *[1 mark]* / excess ions *[1 mark]*
   b) E.g. in a healthy/normal kidney, all the glucose is reabsorbed *[1 mark]* so there should be no glucose present in the patient's urine *[1 mark]*.
   c) E.g. the amount needed by the body *[1 mark]*.

Page 43: Controlling Water Content

1 a) pituitary gland *[1 mark]*
   b) If ADH production is reduced, the kidneys will reabsorb less water back into the blood *[1 mark]*, so more water will be lost from the body in urine *[1 mark]*. If the water that is lost is not replaced/if too much water is lost, dehydration may occur *[1 mark]*.
2 a) As the runner ran, she sweated, resulting in water loss and less water in her blood *[1 mark]*. Her brain detected the decreased water content of her blood *[1 mark]* and instructed the pituitary gland to release ADH/anti-diuretic hormone into the blood *[1 mark]*. The ADH caused the kidneys to reabsorb more water *[1 mark]* resulting in less water being released in her urine, so her urine was more concentrated and appeared darker in colour *[1 mark]*. *[Maximum 4 marks available]*
   b) E.g. it was a cooler day/the runner ran at a slower pace, so she sweated less and lost less water / The runner drank more water during the run, so she replaced the water she lost through sweating *[1 mark]*. This meant that the water content of her blood wasn't as low as it had been the previous day *[1 mark]*. So the pituitary gland released less ADH into the blood *[1 mark]*. This caused less water to be reabsorbed in the kidneys, so her urine was more dilute and paler in colour *[1 mark]*.

Pages 44-45: Kidney Failure

1

Feature of treatment	Dialysis	Transplant
High risk of infection	✓	✓
Long-term, one-off treatment		✓
Patient needs to visit the hospital several times per week	✓	
Patient can lead a relatively normal life		✓

[1 mark for each correctly placed tick]

2 a) They are too large to pass/diffuse through the membrane, so they will remain in the blood *[1 mark]*.
   b) i) It is equal/the same on both sides of the membrane *[1 mark]*. This prevents the diffusion of glucose out of the bloodstream and into the fluid, as it would then be lost from the patient's body *[1 mark]*.
      ii) There is urea in the blood, but not in the dialysis fluid *[1 mark]*. This is so that the harmful urea will diffuse out of the bloodstream and into the fluid, so that it can be removed from the patient's body *[1 mark]*.
   c) To keep the concentrations of dissolved substances in the blood at normal levels *[1 mark]*.
3 a) i) E.g. that there were more people on the waiting list on 31/03/12 than received a transplant over the next year *[1 mark]*. That there were a similar number of people on the transplant list on 31/03/12 and 31/03/13 *[1 mark]*. That some people left the transplant list between 31/03/12 and 31/03/13 *[1 mark]*. *[Maximum of 2 marks available]*
      ii) E.g. there weren't enough donor kidneys available for transplant *[1 mark]*.
   b) $(30\ 800 \times 3) - (17\ 000 + (5000 \times 3)) = \textbf{£60\ 400}$
      [2 marks for correct final answer, otherwise 1 mark for correct working]
      The cost of the transplant is a one-off charge and then after there is just the cost of the drugs every year.

c) E.g. a transplant saves a patient having to make regular, time-consuming trips to the hospital for dialysis sessions *[1 mark]*.

Page 46: Controlling Body Temperature

1 The blood vessels constrict *[1 mark]*. This reduces the blood flow to the surface of the skin *[1 mark]*, so less heat is transferred from the blood to the surroundings *[1 mark]*.
2 a) Skin temperature will increase, so the sweat glands will produce lots of sweat to help cool the body *[1 mark]*. Drinking water will help to replace/balance the water lost in the sweat *[1 mark]*.
   b) The blood vessels widen/dilate *[1 mark]*. This allows more blood to flow near to the surface of the skin *[1 mark]* so more heat can be lost/radiated to the surroundings, cooling the body *[1 mark]*.
3 a) When muscles shiver they contract *[1 mark]*. This contraction needs respiration *[1 mark]*, which releases energy that warms the seal's body *[1 mark]*.
   b) When the seal doesn't shiver, its respiration rate will be lower *[1 mark]* and so it will use less oxygen *[1 mark]*. This will increase the amount of time it can spend underwater *[1 mark]*.

Pages 47-48: Controlling Blood Glucose

1 a) pancreas *[1 mark]*
   b) insulin *[1 mark]*, pancreas *[1 mark]*, cells/tissues *[1 mark]*, reduced *[1 mark]*
   c) The hormone glucagon is secreted by the pancreas *[1 mark]*. This causes glycogen to be converted into glucose *[1 mark]*, which is released into the blood, causing the blood glucose level to increase *[1 mark]*.
      Be careful when you're spelling 'glucagon' and 'glycogen' — they're very similar sounding words and it's easy to get them mixed up. You need to make sure you get them spot on in the exam or the examiner won't know which one you mean and you could lose marks.
   d) When glycogen stores (in the liver and muscles) are full *[1 mark]*.
2 a) in the muscles *[1 mark]*
   b) Eating releases glucose into the bloodstream *[1 mark]*. Excess glucose cannot be converted into glycogen since there's not enough functioning glycogen synthase *[1 mark]*, so the blood glucose level increases *[1 mark]*.
   c) insulin *[1 mark]*
3 a) He needs to inject insulin to lower his blood glucose level *[1 mark]* because his pancreas does not produce enough insulin to do this effectively *[1 mark]*.
   b) 12:30 *[1 mark]* because after this time there is a sharp decrease in the blood glucose level *[1 mark]*.
   c) i) Her body doesn't respond to its own insulin/is resistant to its own insulin *[1 mark]* so injecting extra/additional insulin wouldn't have any effect on her blood glucose levels *[1 mark]*. She could take drugs to improve the way her cells respond to insulin instead *[1 mark]*.
      ii) E.g. obesity *[1 mark]*
   d) E.g. careful diet *[1 mark]* / (regular) exercise *[1 mark]*

Section Five — Defending Against Disease

Page 49: Fighting Disease

1 a) E.g. by producing toxins (poisons) *[1 mark]*. / By damaging your cells *[1 mark]*.

b) Viruses invade body cells and replicate inside the cells *[1 mark]*. This damages the cells, making you feel ill *[1 mark]*.

2 a) A pathogen is a microorganism that causes disease *[1 mark]*.

b) i) antibodies *[1 mark]*, pathogens *[1 mark]*, particular *[1 mark]*, ingesting *[1 mark]*.

ii) Antitoxins counteract toxins produced by bacteria *[1 mark]*.

c) If a person is infected with chickenpox again, their white blood cells will rapidly produce the antibodies to kill the virus *[1 mark]*. So they are naturally immune to the virus and won't get ill *[1 mark]*.

Page 50: Fighting Disease — Vaccination

1 a) i) immune *[1 mark]*

ii) mumps, rubella *[1 mark]*

b) When child A was vaccinated, he was injected with small amounts of dead or inactive pathogens (including the measles pathogen) *[1 mark]*. Child A's white blood cells produced antibodies to attack these pathogens *[1 mark]*. When child A comes into contact with the measles pathogen again, his white blood cells rapidly produce antibodies to destroy the pathogen *[1 mark]*. Child B isn't able to produce antibodies to destroy the measles pathogen as quickly *[1 mark]*. Child A is immune to measles, child B isn't *[1 mark]*.
[Maximum of 4 marks available]

c) E.g. the risk of side effects is small compared to the risk of catching one of the MMR diseases *[1 mark]*. If a large proportion of the population is vaccinated, then none of the diseases the MMR protects against will spread as much because there are fewer people who can pass them on/a large outbreak of any of these diseases can be prevented *[1 mark]*.

Pages 51-52: Fighting Disease — Drugs

1 a) The painkillers will not cure the cold as they do not kill the pathogens that cause it *[1 mark]*. They will help to reduce the symptoms of the cold *[1 mark]*.

b) i) E.g. penicillin *[1 mark]*

ii) If the cold is caused by a virus, the antibiotics won't have any effect *[1 mark]* as antibiotics only destroy bacteria *[1 mark]*.

2 a) 9 days *[1 mark]*

b) Different antibiotics kill different types of bacteria, so it's important that the patient is treated with the right one *[1 mark]*.

c) E.g. viruses reproduce using human body cells *[1 mark]* so it is difficult to develop drugs that destroy the virus without killing the body's cells *[1 mark]*.

3 a) The number of deaths has decreased *[1 mark]*.

b) i) E.g. by overusing/over-prescribing antibiotics *[1 mark]*. By using antibiotics inappropriately *[1 mark]*.

ii) E.g. drug companies are developing new antibiotics that are effective against the resistant strains *[1 mark]*. Doctors are not usually prescribing antibiotics for mild infections (such as a sore throat) / infections that are not serious *[1 mark]*.

4 a) People in hospitals are more likely to have a weakened immune system *[1 mark]* so are more likely to become seriously ill if they are exposed to MRSA *[1 mark]*.

b) i) *Staphylococcus aureus* bacteria mutated *[1 mark]* causing them to be resistant to/not killed by the antibiotic methicillin *[1 mark]*. When people with a *Staphylococcus aureus* infection were treated with methicillin, only the non-resistant bacteria were killed *[1 mark]*. The methicillin-resistant bacteria survived and reproduced, increasing the population of MRSA bacteria *[1 mark]*.

ii) People would not be immune to the new strain of bacteria *[1 mark]* and there may be no effective treatment for it *[1 mark]*, which could allow it to spread rapidly in a population, leading to a big outbreak of disease *[1 mark]*.

Page 53: Investigating Antibiotic Action

1 a) How to grade your answer:

0 marks: There is no relevant information.

1-2 marks: There is a basic method that shows some understanding of how to investigate the action of disinfectant on bacteria.

3-4 marks: There is a clear method that would allow valid results to be collected, including some detail on how to make the investigation a fair test and how to ensure the investigation is carried out safely.

5-6 marks: There is a clear and detailed method that would allow valid results to be collected, including how to ensure that the investigation is a fair test and carried out safely. The answer has a logical structure and uses correct spelling, grammar and punctuation.

Here are some points your answer may include:
Sterilise a Petri dish and some agar jelly.
Pour the agar jelly into the Petri dish and leave it to set.
Sterilise an inoculating loop by passing it through a (Bunsen burner) flame.
Make sure the Bunsen burner is on a heatproof mat.
Use the inoculating loop to transfer microorganisms from the work surface to the agar jelly.
Only lift the lid of the Petri dish a small amount when transferring the microorganisms, to prevent contamination.
Soak five paper discs in the five different disinfectants and place them on the agar jelly.
Place one disc that has not been soaked in disinfectant on the agar jelly. This is the control.
All the paper discs should be the same size.
Use tape to secure the lid of the Petri dish to stop other microorganisms from getting in.
Place the Petri dish in an incubator.
Make sure the incubator is at about 25 °C to prevent harmful microorganisms growing.
Store the Petri dish upside down to stop bacteria falling onto the agar surface.
After a set period of time, remove the Petri dishes from the incubator. Measure the diameter of the clear zone around the disc.
The disc soaked in the most effective disinfectant will have the widest clear zone (area where no bacteria are growing).

You don't need to make all of these points to get full marks, but you do need to make sure you've covered all the things you're told to write about in the question. In this case that's:
* *the method and equipment the student would use.*
* *how the student would make the investigation a fair test.*
* *any safety precautions the student should take.*

b) Because in industrial conditions bacteria are incubated at higher temperatures than in a school laboratory *[1 mark]*, so they can grow a lot faster *[1 mark]*.

Section Six — Plants

Page 54: Photosynthesis

1 a) i) water *[1 mark]*
   ii) oxygen *[1 mark]*
  b)
$$6CO_2 + 6H_2O \xrightarrow{\text{light energy}} C_6H_{12}O_6 + 6O_2$$
   [1 mark for 6CO₂ and 6H₂O on left-hand side of equation, 1 mark for C₂H₁₂O₆ and 6O₂ on right-hand side of equation]
  c) the air *[1 mark]*
2 a) i) Z *[1 mark]*
   ii) chloroplast *[1 mark]*
  b) Chlorophyll absorbs the light energy needed for photosynthesis to take place *[1 mark]*.

Pages 55-56: The Rate of Photosynthesis

1   The temperature will be lower in winter than in summer *[1 mark]* and there will be less (day)light/the light intensity will be lower in winter than in summer *[1 mark]*. Both of these factors will reduce the rate of photosynthesis *[1 mark]* and therefore the growth rate of the grass *[1 mark]*.

2 a) At low light intensities, increasing the CO_2 concentration has no effect *[1 mark]*, but at higher light intensities, increasing the concentration of CO_2 increases the maximum rate of photosynthesis *[1 mark]*.
  b) The rate of photosynthesis does not continue to increase because temperature or the level of carbon dioxide becomes the limiting factor *[1 mark]*.

3 a) blue light, approx. 440 nm *[1 mark]*,
      red light, approx. 660 nm *[1 mark]*
  b) E.g. use blue or red light bulbs to increase the rate of photosynthesis, and therefore the growth rate *[1 mark]*.

4   Paraffin heaters increase the carbon dioxide concentration *[1 mark]* and the temperature *[1 mark]* inside a greenhouse. These can both increase the rate of photosynthesis *[1 mark]*, meaning plants will grow faster and bigger, so the crop can be harvested more often/a bigger yield can be collected at each harvest *[1 mark]*.

5 a) the rate of photosynthesis *[1 mark]*
  b)

(Graph: Y axis "Volume of oxygen evolved in 10 minutes (cm³)" from 0 to 80; X axis "Relative light intensity" from 0 to 12; points plotted at approx (2, 12), (4, 25), (6, 15), (8, 47), (10, 61), (12, 72))

   [1 mark for correctly labelling the Y axis as volume of oxygen evolved in 10 minutes (including use of the correct units), 2 marks for correctly plotted points. Deduct up to 2 marks for incorrectly plotted points, 1 mark per incorrect point.]
  c) 6 *[1 mark]* E.g. the student might have measured the volume of oxygen evolved in less time than 10 minutes *[1 mark]*. / The student might have accidentally used a lower light intensity *[1 mark]*. / The student might have measured the volume of oxygen wrongly *[1 mark]*.
  d) The relationship would continue up to a point, and then the graph would level off *[1 mark]*. At this point, the temperature, carbon dioxide level or both of these factors would be acting to limit the rate of photosynthesis *[1 mark]*.

Page 57: How Plants Use Glucose

1 a) respiration *[1 mark]*, cellulose *[1 mark]*
  b) i) E.g. fat(s)/oil(s)/lipid(s) *[1 mark]*
   ii) E.g. when photosynthesis isn't happening/at night *[1 mark]* / when the rate of photosynthesis is low/in winter *[1 mark]*
  c) E.g. because starch is insoluble, unlike glucose *[1 mark]*. This means that starch doesn't cause water to be drawn into cells (via osmosis) and swell up *[1 mark]*.
2 a) i) through photosynthesis *[1 mark]*
   ii) Sundews can absorb nitrate ions from the insects they catch *[1 mark]*. Most plants can't do this — they absorb nitrate ions from the soil instead *[1 mark]*. But carnivorous plants like sundews are not able to absorb enough ions in this way because they grow on nutrient-poor soils *[1 mark]*.
Carnivorous plants are carnivorous __because__ they grow on nutrient-poor soils — they don't get enough nutrients from the soil, so they get them from the animals they catch instead.
  b) E.g. Venus Fly Trap *[1 mark]*

Pages 58-59: Exchanging Materials in Plants

1 a) i) E.g. water *[1 mark]*, mineral ions *[1 mark]*
   ii) E.g. their surface is covered in root hairs *[1 mark]*, which increase the surface area of the roots for absorption *[1 mark]*.
  b) E.g. leaves have a flattened shape *[1 mark]* and internal air spaces *[1 mark]*.
2 a) diffusion *[1 mark]*
  b) i)

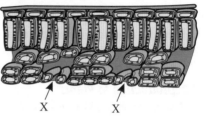

   X   X

   [1 mark for correctly placing a label in either of the locations shown above]
   ii) oxygen *[1 mark]*.
  c) The guard cells are unable to close the stomata *[1 mark]*. The mutant plants wilt in the hot, dry environment, which means they must have lost water through their stomata faster than it can be replaced by the roots *[1 mark]*. In the non-mutant plants, the stomata are closed by the guard cells to prevent this from happening *[1 mark]*.
3 a) 9 a.m. *[1 mark]*
  b) Any two from: e.g. day 2 was colder *[1 mark]*. / Day 2 was less windy *[1 mark]*. / Day 2 was wetter/more humid *[1 mark]*.
  c) The plant has lost too much water/has lost water faster than it could be replaced through the roots *[1 mark]*.
  d) The rate of water vapour loss would be slower *[1 mark]* since most water loss occurs through the stomata *[1 mark]*, which are found mainly on the underside of the leaves *[1 mark]*.

Page 60: Water Flow Through Plants

1 a) Water has evaporated from the stomata of the flower petals, creating slight shortage of water in the petals *[1 mark]*. This has caused more water to be transported up through the xylem of the cut stem in the transpiration stream *[1 mark]* and the dye has been transported along with it *[1 mark]*.
  b) i) E.g. you could time exactly how long it takes for the dye to reach the flower at a range of different temperatures *[1 mark]*.
   ii) Any two from: e.g. the amount of dye in the beaker *[1 mark]* / the length of the stem *[1 mark]* / the thickness of the stem *[1 mark]* / the size of the flower *[1 mark]*.

2   The removal of the phloem means that the tree will no longer be able to transport sugars from the leaves to the growing regions/storage organs / translocation will be prevented *[1 mark]*. This will stop the roots from getting food / may stop the tree from growing/producing new shoots/storing food *[1 mark]* which could eventually kill it *[1 mark]*.

Pages 61-62: Plant Hormones

1   light *[1 mark]*, phototropism *[1 mark]*, gravitropism/geotropism *[1 mark]*, moisture/water *[1 mark]*

2 a)   auxin *[1 mark]*

  b) i) The root will grow down again *[1 mark]*. Gravity/moisture in the soil will cause auxin to build up on the lower side of the root *[1 mark]* inhibiting growth of the lower cells, so the root will bend downwards *[1 mark]*.

    ii) E.g. this response means that a plant's roots are more likely to find nutrients/moisture in the soil, which are/is essential for plant survival *[1 mark]*.

  c) E.g. as weedkillers *[1 mark]*, as rooting powders/hormones *[1 mark]*

3   Seedling A: e.g. the foil prevents any light reaching the tip *[1 mark]*, so the auxin is remains evenly distributed in the tip and no bending occurs *[1 mark]*.
Seedling C: e.g. the mica strip prevents the auxin from moving to the side of the shoot in the shade *[1 mark]*, so there is an even distribution of auxin and no bending occurs *[1 mark]*.

4 a)   0.001 (parts per million) *[1 mark]*

  b) E.g. the highest auxin concentration appears to inhibit the rate of growth *[1 mark]*, since the increase in root length is less than when the auxin concentration is zero *[1 mark]*.

Page 63: Sexual Reproduction in Plants

1   D, A, C, B *[2 marks for all steps in correct order, otherwise 1 mark for at least two steps in the correct order]*

2 a) i) gametes *[1 mark]*, stigma *[1 mark]*, style *[1 mark]*

    ii) E.g. a nucleus from the male gamete moves down the pollen tube into the ovary *[1 mark]* where it fertilises the egg cell nucleus in the ovule *[1 mark]*. This forms a zygote, which develops into the embryo *[1 mark]*. Other nuclei from the pollen grain fertilise endosperm nuclei in the ovule *[1 mark]*. This forms the endosperm, which along with the female tissues in the ovule make up a seed *[1 mark]*.
[Maximum of 4 marks available]

  b) i) anther *[1 mark]*

    ii) ovary *[1 mark]*

    iii) style *[1 mark]*

If you put carpel as the answer to part iii) then you're technically correct, but the arrow is pointing to the specific part of the carpel called the style — so that's the answer that will get you the mark.

Section Seven — Variation and Genetics

Page 64: Variation

1 a)   The difference in weight must be caused by the environment *[1 mark]*, because the twins have exactly the same genes *[1 mark]*.

In this case, the environment can include the amount of food each twin eats and/or the amount of exercise they each do.

  b) No, because if they were caused by genes both twins would have the birthmark *[1 mark]*.

2 a)   E.g. the foal will inherit genes from both its parents *[1 mark]*. The genes carry information that leads to the foal developing similar characteristics to its parents *[1 mark]*.

  b) There is no guarantee that the foal will be a successful racehorse because the combining of genes from two parents causes genetic variation *[1 mark]*. The particular combination of genes the foal inherits might not make it genetically suited to racing *[1 mark]*. In addition, environmental conditions may contribute to how successful the foal is and these cannot be exactly replicated/completely controlled *[1 mark]*.

Page 65: The Structure of DNA

1 a)   E.g.

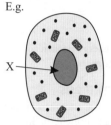

[1 mark for correctly labelling the nucleus with a letter X]

  b) two *[1 mark]*

  c) i) deoxyribonucleic acid *[1 mark]*

    ii) E.g. it contains the coded information that determines inherited characteristics *[1 mark]*.

    iii) A DNA molecule is made up of (two) long strands that are coiled together in a double helix *[1 mark]*. The strands are held together by four different chemicals called bases *[1 mark]*.

2 a)   E.g. a sequence of three bases in a gene codes for a particular amino acid *[1 mark]*. Each protein is made up of a specific sequence of amino acids *[1 mark]*. The order of amino acids in a protein is determined by the order of the bases in the gene *[1 mark]*.

  b) E.g. a different protein might be produced/a protein might not be produced at all/the p53 protein produced might not function properly *[1 mark]*. This could prevent cell division being properly regulated and lead to the growth of tumours *[1 mark]*.

Page 66: X and Y Chromosomes

1 a) i) E.g. sex is determined by two chromosomes/a single pair of chromosomes in both chickens and humans *[1 mark]*.

    ii) E.g. in humans, the sex chromosomes are called X and Y, not Z and W *[1 mark]*. In humans, it is the female that has two of the same sex chromosomes (XX)/the male that has two different sex chromosomes (XY) — in chickens, it's the opposite way round *[1 mark]*.

  b) 23 *[1 mark]*

2 a)   E.g.

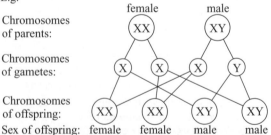

[1 mark for correct chromosomes in parents, 1 mark for correct chromosomes in gametes, 1 mark for correct chromosomes in offspring, 1 mark for correct sex of offspring]

  b) Male children will not inherit the colour blindness allele because they only inherit a Y chromosome from their father *[1 mark]*.

Page 67: The Work of Mendel

1 a) i) E.g. inherited factors determine the characteristics of an organism *[1 mark]*. They are passed from parents to offspring/generation to generation *[1 mark]*. The modern word for them is genes *[1 mark]*. *[Maximum of 2 marks available]*

ii) the dominant 'inherited factor' *[1 mark]*

b)

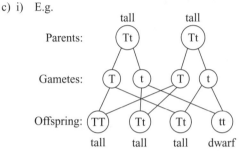

[1 mark for correctly showing the gametes' alleles, 1 mark for correctly showing the offspring's alleles, 1 mark for correctly stating that all the offspring would be tall]

c) i) E.g.

[1 mark for correctly showing the gametes' alleles, 1 mark for correctly showing the offspring's alleles, 1 mark for correctly stating that three offspring would be tall and one would be dwarf]

ii) 75% / 0.75 / ¾ *[1 mark]*

Page 68: Genetic Diagrams

1 a) An allele is a particular form/version of a gene *[1 mark]*.

b) i) heterozygous *[1 mark]*

ii) brown hair *[1 mark]*

2 a) E.g.

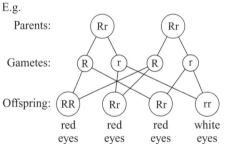

[1 mark for correct genotypes of the parents, 1 mark for correct gamete genotypes, 1 mark for correct offspring genotypes, 1 mark for correct offspring phenotypes]

It's always worth having a go at drawing a genetic cross in the exam, as there are so many marks available for them. If you'd made a mistake here and got the parental genotypes wrong, you could still have picked up some marks for following the cross all the way through to the end.

b) i) 25% / 0.25/ 1 in 4 / ¼ *[1 mark]*

ii) 75% of the offspring are likely to have red eyes (75 ÷ 100) × 60 = **45** *[2 marks for the correct answer, otherwise 1 mark for the correct working]*

Page 69: More Genetic Diagrams

1 a) homozygous *[1 mark]*

b) Parent genotype: Aa *[1 mark]*
Gamete genotypes: A, a *[1 mark]*

c) i) 50% / 0.5 / 1 in 2 / ½ *[1 mark]*

ii) Fertilisation is random/the genetic diagram only shows the probability of the outcome, so the numbers of offspring produced will not always be exactly in those proportions *[1 mark]*.

2 Hh *[1 mark]*. Huntington's disease is a dominant disorder, so if she was HH all of her children would be affected (and Brook is not) *[1 mark]*.

3 The gardener can cross the plants *[1 mark]*. If the parent plant with red flowers is thoroughbred (RR) then all the offspring will be red *[1 mark]*. If the parent plant with red flowers is hybrid (Rr) then there will be an approximate 1:1 ratio of red to white offspring *[1 mark]*.

Pages 70-71: Genetic Disorders

1 E.g. the presence of an extra chromosome/an abnormal number of chromosomes *[1 mark]*.

2 a) having extra fingers or toes *[1 mark]*

b) the female from Couple A *[1 mark]*

c) i) E.g.

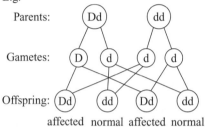

[1 mark for correct genotypes of the parents, 1 mark for correct gamete genotypes, 1 mark for correct offspring genotypes, 1 mark for correct offspring phenotypes]

ii) 50% / 0.5 / 1 in 2 / ½ *[1 mark]*

3 a) i)

	Wife's alleles	
	F	F
F	FF	FF
f	Ff	Ff

(Husband's alleles)

[1 mark for four correct alleles for the husband and wife, 1 mark for four correct offspring genotypes]

Make sure you write your letters clearly in this answer. In the exam the examiner needs to be able to tell the difference between lower case letters and capitals. If there isn't a clear difference between them you might lose marks.

ii) 0 *[1 mark]*

b) No. Cystic fibrosis is a recessive disorder, so to have it you must inherit a copy of the faulty allele from both parents *[1 mark]*. This isn't possible because the brother's wife doesn't have the faulty allele *[1 mark]*.

4 a) red blood *[1 mark]*.

b) Heterozygous individuals are less likely to get malaria than people who are homozygous for the dominant allele *[1 mark]*.

c) i) Aa *[1 mark]*

ii) 25% *[1 mark]*

Page 72: Embryo Screening

1 a) A sample of DNA *[1 mark]*.

b) E.g. a gene probe is made that will bind to an allele for a specific disorder *[1 mark]*. The probe is then mixed with a sample of DNA from the embryo — if the allele is present, the probe will bind to it *[1 mark]*. UV light can then be shone on the DNA sample to see if the probe has bound to the allele (if the probe has attached then the sample will glow) *[1 mark]*.

152

c) How to grade your answer:
0 marks: No relevant information is given.
1-2 marks: A brief mention is made of one argument for and one against embryo screening.
3-4 marks: At least two arguments are made for and two against embryo screening. The answer has a logical structure and spelling, grammar and punctuation are mostly correct.
5-6 marks: At least three clear, detailed arguments are made for and three against the use of embryo screening. The answer has a logical structure and uses correct spelling, grammar and punctuation.

Here are some points your answer may include:

For:

For embryos produced by IVF, only the embryos without faulty alleles will be selected for implantation into the womb — this will increase the chance of the mother giving birth to a healthy baby.

Screening embryos helps to stop people suffering from genetic disorders.

Treating genetic disorders can be very expensive, so screening embryos could help to reduce health care costs.

Against:

Taking cells from the womb can increase the chances of a miscarriage, so the process of screening could actually damage a developing embryo.

The results of embryo screening aren't always completely reliable, which could lead to a healthy embryo being terminated because of inaccurate results from the screening. There are different opinions about what counts as a morally acceptable reason to terminate a pregnancy. Some people are worried that embryo screening could be used to have a baby with 'desirable characteristics' and that this is unacceptable.

Embryo screening is an expensive process.

Using embryo screening implies that people with genetic disorders are 'undesirable'. This could increase prejudice towards these people in society.

Page 73: Reproduction

1 a) male and female gametes/sex cells *[1 mark]* fuse together *[1 mark]*
   b) E.g. the offspring inherits a mixture of genes/genetic material from two parents *[1 mark]*.
2 a) clones *[1 mark]*
   b) Asexual reproduction *[1 mark]* because this type of reproduction involves no mixing of genetic material *[1 mark]* and therefore produces no genetic variation in the offspring *[1 mark]*.
   c) E.g. he can be sure of the characteristics of the new plants since they will be identical to those of the original plant(s). / It's a way of helping to make sure that the new potatoes will taste good/be resistant to particular pests *[1 mark]*.

Pages 74-75: Cloning

1 a) A small number of plant cells are put into a growth medium (with hormones), so they grow into new plants *[1 mark]*.
   b) E.g. tissue culture produces many plants that are genetically identical and share the same desirable traits *[1 mark]*. Allowing plants to reproduce sexually could produce plants with undesirable traits, which would be a waste of time/money to grow *[1 mark]*.
2 a) E.g. the embryo that develops is split many times before any cells become specialised *[1 mark]* to form clones/other identical embryos *[1 mark]*. These embryos are then implanted into host mothers to develop fully into cloned calves *[1 mark]*.
   b) i) Because the original embryo was created using sexual reproduction *[1 mark]*, which creates genetic variation in the offspring *[1 mark]*.

ii) E.g. it allows her to produce more calves in a shorter space of time/it allows her to produce lots of calves from her best animals *[1 mark]*.
3 a) A and C *[1 mark]*, because cell C has received the nucleus/genetic material from cell A *[1 mark]*.
   b) X *[1 mark]*. Sheep X provided the genetic material/nucleus for the lamb *[1 mark]*.
   c) an electric shock *[1 mark]*
   d) In sexual reproduction, the sperm and egg each supply half of the genetic material *[1 mark]*. In cloning, a single nucleus provides all the genetic material *[1 mark]*.
4 a) E.g. studying animal clones could help us to understand the development of the embryo/ageing/age-related disorders *[1 mark]*. / Cloning could be used to help preserve endangered species *[1 mark]*.
   b) The animals will all have the same genes *[1 mark]*. So if a new disease appears, there may be no allele in the population giving resistance to it and the animals might all be affected by the disease/wiped out by the disease *[1 mark]*.

Page 76: Genetic Engineering

1 a) E.g. enzymes would be used to isolate the human insulin gene from human DNA *[1 mark]*. The insulin gene would then be inserted into a vector/plasmid *[1 mark]*. This vector/plasmid would be used to insert the insulin gene into a bacterium *[1 mark]*.
   b) viruses *[1 mark]*
2 a) i) early *[1 mark]*.
   ii) E.g. they can be made resistant to herbicides *[1 mark]*.
   b) The GM maize should have a higher yield than the non-GM maize (because it won't be eaten by insect pest) *[1 mark]*.
   c) Any two from: e.g. GM crops might have a negative effect on/reduce wild flower and insect populations *[1 mark]*. / GM crops might negatively impact on human health *[1 mark]*. / Transplanted genes may pass into other organisms in the environment *[1 mark]*.
   d) So that the performance of GM and non-GM crops (e.g. their yield) can be compared fairly *[1 mark]*.

Section Eight — Adaptation and Environment

Pages 77-78: Adaptations

1 extremophiles *[1 mark]*, temperature/pressure *[1 mark]*, pressure/temperature *[1 mark]*
2 a) In the hair/fur of mammals *[1 mark]*.
   b) E.g. help it suck blood *[1 mark]*, it is not easily brushed off (the animal) *[1 mark]*, it is not damaged *[1 mark]*
   c) E.g. it enables the flea to jump between hosts *[1 mark]*.
3 a) E.g. they have suckers and hooks *[1 mark]*.
   b) To produce large numbers of eggs *[1 mark]*.
   c) E.g. they have a thick outer cuticle *[1 mark]* to protect them from the host's digestive enzymes *[1 mark]*. They have a flattened shape *[1 mark]*, which provides a large surface area for the absorption of soluble food from the host's gut *[1 mark]*.
4 a) Any two from: e.g. they help the bears to climb trees to get food. / They help the bears to climb trees to escape predators. / They help the bears to catch food. / They help the bears to defend themselves against predators/other animals. / They are used to fight other bears. *[2 marks available, 1 mark for each correct answer]*

Answers

b) i) E.g. lowering their metabolism conserves energy *[1 mark]* so they don't have to hunt when it's cold/there's not much food about *[1 mark]*.

ii) E.g. it allows them to build up their fat reserves/put on weight before winter, when food may be scarce *[1 mark]*.

Page 79: The Malaria Parasite

1 a) The gametocytes reproduce sexually inside the mosquito *[1 mark]*. They develop into sporozoites in the mosquito's salivary glands *[1 mark]*.

b) How to grade your answer:
0 marks: No relevant information is given.
1-2 marks: There is a basic description of the life cycle of the malaria parasite.
3-4 marks: There is a description of the life cycle of the malaria parasite, which mentions merozoites, schizonts or red blood cells. The answer has a logical structure and spelling, grammar and punctuation are mostly correct.
5-6 marks: There is a clear, detailed description of the life cycle of the malaria parasite, which mentions merozoites, schizonts, the liver and red blood cells. The answer has a logical structure and uses correct spelling, grammar and punctuation.

Here are some points your answer may include:
The sporozoites travel in the bloodstream to the liver.
In the liver cells, the sporozoites divide asexually to form merozoites.
Thousands of merozoites are released from liver cells into the blood, where they invade red blood cells.
Some merozoites grow inside the red blood cells and their nuclei divide asexually to produce cells with many nuclei called schizonts.
The schizonts then divide to produce many merozoites.
The red blood cells burst, releasing merozoites into the blood.
Toxins are also released, triggering fever attacks.
The released merozoites then go on to invade other red blood cells.
Some merozoites enter a sexual phase of reproduction to produce gametocytes.

c) Each form is adapted for living in a different place in the animal or mosquito host *[1 mark]*.

Pages 80-81: Competition and Environmental Change

1 a) The two species would have to compete for it *[1 mark]*.
b) E.g. as a source of food *[1 mark]*.
c) i) Any two from: e.g. territory / mates / food
[2 marks available, 1 for each correct answer]
ii) Any two from: e.g. light / nutrients/minerals from the soil / space / water *[2 marks available, 1 for each correct answer]*

2 a) The maximum altitude/height up the mountain where the snail was found increased *[1 mark]*.
b) E.g. the snail species may have been found higher up the mountain because it became warmer than it used to be higher up *[1 mark]*.
c) E.g. a change in average rainfall / a change in the availability of oxygen/water/carbon dioxide/light/habitat/shelter *[1 mark]*.
d) E.g. bees (in the US) *[1 mark]* / the European Bee-Eater bird *[1 mark]* / some bird species *[1 mark]*

3 a) Any two from: e.g. the temperature may have increased. / The amount of light may have increased. / The level of nutrients may have increased. / The numbers of competitors may have decreased. / The number of fish/predators may have decreased. *[2 marks available, 1 for each correct answer]*
b) May (also accept April) *[1 mark]*. The number of algae decreased sharply in May *[1 mark]*. This could have been caused by an increase in the number of fish *[1 mark]*.

4 a) E.g. as the number of foxes increases, the number of rabbits decreases *[1 mark]* because more rabbits are eaten by the foxes *[1 mark]*. This eventually means that there is less food for the foxes *[1 mark]* and so the number of foxes decreases *[1 mark]*. This allows the number of rabbits to increase again, as there is less predation *[1 mark]*, and so on. *[Maximum of 4 marks available]*
b) E.g. mates / territory *[1 mark]*.

Page 82: Measuring Environmental Change

1 a) E.g. the more polluted the water, the lower the number of mayfly larvae *[1 mark]* and the higher the number of sludgeworms *[1 mark]*. / The less polluted the water, the higher the number of mayfly larvae *[1 mark]* and the lower the number of sludgeworms *[1 mark]*.
b) The sample taken at 3 km *[1 mark]*.
Pollution of water by sewage reduces the concentration of dissolved oxygen in the water.

2 a) E.g. lichens *[1 mark]*
b) E.g. using specialist equipment allowed the exact sulfur dioxide concentration to be measured, which couldn't have been done with a living indicator *[1 mark]*. However, the equipment only gave a snapshot of the conditions at the time the measurements were taken, whereas using a living indicator would have given a more long-term picture *[1 mark]*. Using specialist equipment may have also been more expensive/taken more specialist training than using a living indicator *[1 mark]*.

Pages 83-84: Monitoring the Distribution of Organisms

1 a) E.g. he could have divided the field into a grid and used a random number generator to select coordinates *[1 mark]*.
b) i) Mean number of daisies = $(3 + 1 + 2 + 1 + 4 + 3 + 0 + 2) \div 8$ = **2** per quadrat
[2 marks for correct answer, otherwise 1 mark for correct working]
ii) Number of daisies per quadrat in size order = 0, 1, 1, 2, 2, 3, 3 4, so middle value (median) is **2** per quadrat. *[2 marks for correct answer, otherwise 1 mark for arranging the values for the number of daises per quadrat in size order]*
iii) 2 daisies per quadrat = 2 daisies per 1 m^2
Field = 5600 m^2 so total population of daisies = 5600×2 = **11 200**
[2 marks for correct answer, otherwise 1 mark for correct working. Award marks if incorrect answer to b) i) is used correctly here.]
c) E.g. by using a larger sample size/data from more quadrats *[1 mark]*.

2 a) E.g. he could have set up a transect from the wood to the opposite side of the field *[1 mark]*, then placed 1 m^2 quadrats at regular/2 metre intervals along the transect *[1 mark]*. Then he could have counted the dandelions in each quadrat *[1 mark]*.
b) The number of dandelions increases with distance from the wood *[1 mark]*.
c) The results are not valid because they don't answer the original question *[1 mark]*. The student hasn't controlled all of the variables *[1 mark]*, so the investigation is not a fair test/doesn't prove that light intensity is the only variable affecting the dandelions' distribution *[1 mark]*.
d) E.g. different soil moisture levels caused by the stream between the field and the wood *[1 mark]*.

Section Nine — Environment, Evolution and Energy Transfer

Page 85: Human Impact on the Environment

1 a) E.g. the population size increased slowly between 1000 and 1800, from 0.3 billion to 1 billion *[1 mark]*. Between 1800 and the present day the population has increased rapidly from 1 billion to over 7 billion *[1 mark]*.

b) Any three of: e.g. more land needs to be farmed *[1 mark]* / more land is used for building *[1 mark]* / more land is taken up for dumping waste *[1 mark]* / more land is taken up by quarrying *[1 mark]*.

c) Any two of: e.g. sewage *[1 mark]* pollutes water/causes eutrophication *[1 mark]*. / Smoke/gases *[1 mark]* cause air pollution *[1 mark]*. / Sulfur dioxide *[1 mark]* causes air pollution/acid rain *[1 mark]*. / Fertilisers *[1 mark]* pollute water/cause eutrophication *[1 mark]*. Toxic chemicals/pesticides/herbicides *[1 mark]* pollute water/land *[1 mark]*.

d) Because the average standard of living has also increased (meaning that the average person consumes more and makes more waste) *[1 mark]*.

Page 86: Eutrophication

1 a) As the nitrate concentration increased, the average number of fish per cubic metre decreased *[1 mark]*.

b) E.g. increased use of fertilisers. / Increased runoff/leaching of fertilisers due to higher rainfall *[1 mark]*.

c) Increased levels of nitrates/mineral ions in the water cause algae and/or plants to grow more quickly *[1 mark]*, blocking out the light and preventing organisms below from photosynthesising *[1 mark]*. The algae and/or plants start to die, providing food for microorganisms *[1 mark]* which increase in number and use up all the oxygen in the water as they respire *[1 mark]*. Fish need oxygen so, as less oxygen is available, the fish start to die and their numbers decrease *[1 mark]*.

2 The number of microorganisms increases downstream of the sewage pipe *[1 mark]*. The sewage provides extra mineral ions causing algae and/or plants to grow rapidly *[1 mark]*. These algae and/or plants block the light from other plants, stopping them from photosynthesising and causing them to die *[1 mark]*. The dead plants provide food for microorganisms, causing the number of microorganisms to increase *[1 mark]*.

Page 87: Deforestation and Destruction of Peat Bogs

1 a) Any two of: e.g. to clear land for growing rice *[1 mark]*. / To clear land for raising cattle *[1 mark]*. / To make space for growing crops from which biofuels can be made *[1 mark]*.

b) i) Deforestation reduces the amount of carbon dioxide removed from the atmosphere *[1 mark]* because it reduces the number of trees, which remove carbon dioxide via photosynthesis *[1 mark]*.

ii) Trees release carbon dioxide when they are burnt *[1 mark]*. Microorganisms feeding on dead wood release carbon dioxide (when they respire) *[1 mark]*.

c) E.g. the variety of species living in a habitat *[1 mark]*.

2 a) Plants living in bogs do not fully decompose when they die because there is not enough oxygen available *[1 mark]*. This causes a build-up of partially decayed plant material, which forms peat *[1 mark]*.

b) Peat acts as a carbon store *[1 mark]*. As the land is drained, the peat is able to start decaying, releasing carbon dioxide into the atmosphere *[1 mark]*.

c) E.g. by only buying peat-free compost *[1 mark]*.

One reason why peat bogs are drained is so that peat can be sold to gardeners as compost, so if gardeners only buy peat-free compost this will reduce the demand for peat.

Page 88: Carbon Dioxide and the Greenhouse Effect

1 a) i) It has increased the Earth's temperature *[1 mark]* because carbon dioxide is a greenhouse gas / it re-radiates heat back towards the Earth/traps heat close to the Earth *[1 mark]*.

ii) e.g. methane *[1 mark]*

b) i) It removes carbon dioxide from the atmosphere (reducing the impact of global warming) *[1 mark]*.

ii) Any two of: e.g. lakes *[1 mark]* / ponds *[1 mark]* / oceans *[1 mark]* / peat bogs *[1 mark]*

2 a) A1FI *[1 mark]* because the global temperature increases the most in this scenario *[1 mark]*, and this is likely to be caused by the release of more (greenhouse) gases, such as carbon dioxide *[1 mark]*.

b) E.g. they may cause other big changes in the Earth's climate. / They may cause the sea level to rise. / They may cause changes in the migration patterns of some species. / They may cause changes to the distribution of many species *[1 mark]*.

Page 89: Climate Change

1 E.g. sea turtles are likely to become less widely distributed *[1 mark]*. This is because the increased temperature will cause sea levels to rise *[1 mark]*, possibly reducing the number of available beaches for the turtles to lay their eggs on *[1 mark]*.

Low-lying beaches will disappear if sea levels rise.

2 a) As the average April temperature increases, the first day that the swallow is seen in Britain gets earlier *[1 mark]*.

b) Valid data is repeatable, reproducible and answers the original question *[1 mark]*. There is evidence to suggest that the results of both studies are likely to be reproducible as they were both conducted over long time periods *[1 mark]* and both studies found similar results *[1 mark]*. However, more information would be needed about other factors that may be affecting the swallows' migration patterns to prove that global warming is causing the changes seen *[1 mark]*, so the data doesn't answer the original question *[1 mark]*.

[Maximum of 4 marks available.]

Page 90: Evolution

1 a) Sharks *[1 mark]*, because rays and sharks share a more recent common ancestor than rays and dolphins or mice *[1 mark]*.

b) E.g. they are competitors *[1 mark]*.

c) E.g. to classify organisms into groups (such as plants, animals and microorganisms) *[1 mark]*.

2 a) 3 billion *[1 mark]*

b) Originally there was variation in the length of horses' legs (due to genetic differences) *[1 mark]*. The horses with longer legs may have been better able to run away from predators, so were more likely to survive and breed *[1 mark]*. They would have passed the genes for longer legs on to the next generation, so their offspring would also have had longer legs *[1 mark]*. Across many generations, this would have led the average length of horses' legs to increase *[1 mark]*.

Page 91: More About Evolution

1 a) Darwin's theory went against common religious beliefs about how life on Earth had developed. / Darwin's theory challenged the idea that life on Earth needed a creator/God *[1 mark]*. Darwin couldn't explain how new, useful characteristics appeared or were passed onto the next generation, as genes hadn't been discovered *[1 mark]*. There wasn't enough evidence to convince other scientists, as few studies had been carried out *[1 mark]*.

b) i) developed *[1 mark]*, longer *[1 mark]*, similar *[1 mark]*

ii) Because there's not enough evidence to support the hypothesis *[1 mark]*.

2 No, because the useful characteristic that developed during the pigeon's lifetime (a strong left leg) has not been passed on to its offspring *[1 mark]*.

Pages 92-93: Fossils

1 a) Fossils are the remains/imprints of animals or plants that lived many thousands of years ago and have been preserved in rock *[1 mark]*.
 b) i) The dinosaur may have stepped in a soft material (e.g. clay), making an imprint *[1 mark]*. When the material hardened, the imprint was preserved *[1 mark]*.
  ii) The bones in the dinosaur's foot wouldn't have easily decayed, so they would have lasted a long time when buried *[1 mark]*. The bones could eventually have been replaced by minerals as they decayed, forming a rock-like substance shaped like the bones of the dinosaur's foot *[1 mark]*.
 c) Because scientists haven't found many fossils of very early life forms *[1 mark]*. This is because most early life forms were soft-bodied and soft-bodied life forms are more likely to decay without forming fossils *[1 mark]*. Also any fossils that did form are likely to have been destroyed by geological activity *[1 mark]*.
2 a) i) X, because it's closer to the surface, suggesting it was buried the most recently *[1 mark]*.
  ii) Sea creatures with shells will have more hard parts than most plants *[1 mark]*. These hard parts decay more slowly so are more likely to be fossilised *[1 mark]*.
 b) E.g. so that they can learn how much or how little the organism has changed over time *[1 mark]*.
3 It's likely that the man was buried in the ice shortly after he died, so his body didn't decay *[1 mark]* because the conditions were too cold for decay microbes to work *[1 mark]*.

Page 94: Extinction and Speciation

1 a) Any two of: e.g. the rats could compete with the birds for food *[1 mark]*. / The rats could bring diseases that kill birds *[1 mark]*. / The rats could eat the birds/the birds' eggs *[1 mark]*.
 b) Any two of: e.g. changes in the environment that happen too quickly for the species to adapt *[1 mark]*. / A catastrophic event (such as a volcanic eruption/asteroid collision) happens *[1 mark]*. / A new disease is introduced *[1 mark]*.
2 a) speciation *[1 mark]*
 b) When the river dried up, one group of fish were trapped/isolated in lake A and the other in lake B *[1 mark]*. These two populations showed genetic variation because they had a range of different alleles *[1 mark]*. The environments in lake A and lake B were different, so individuals with alleles for characteristics that made them better adapted to their environment *[1 mark]* gradually became more common due to natural selection *[1 mark]*. Over time, the populations became so different that they could no longer breed with each other, making them separate species *[1 mark]*.

Page 95: Pyramids of Biomass

1 C *[1 mark]*, because the total mass of organisms decreases at each trophic level, which is what is shown in pyramid C *[1 mark]*.
2 a) The concentration of DDT in organisms increases as you go up the stages *[1 mark]*.
 b) 13.8 ÷ 0.04 = **345 times** *[2 marks for correct answer, otherwise 1 mark for correct working]*
 c) Because DDT is stored in the tissues of animals and a pyramid of biomass represents the mass of the living tissues *[1 mark]*.

Pages 96-97: Energy Transfer

1 a) Light energy from the Sun is converted into chemical energy by the grass during photosynthesis *[1 mark]*. This energy is stored in the grass cells *[1 mark]*. When the rabbit eats the grass, it obtains this energy *[1 mark]*.
 b) E.g. some of the light that reaches the grass is not absorbed / is used heating up the grass / is absorbed by parts of the grass that do not photosynthesise *[1 mark]*.
 c) 2070 ÷ 10 = 207 kJ available to the second trophic level *[1 mark]*
  207 − (90 + 100) = **17 kJ** available to Animal A *[1 mark]*
 d) Some of the rabbits' energy is lost during life processes, such as movement *[1 mark]* and some energy is lost to the surroundings as heat *[1 mark]*. Biomass and energy are also lost in the rabbits' waste / in inedible parts of the rabbits *[1 mark]*.
2 C *[1 mark]*. This is the shortest food chain. As energy is lost at each stage of a food chain, the shortest will be most efficient *[1 mark]*.
3 a) C, because at this farm the birds have the least space available, so they will have the most limited movement *[1 mark]* and they are kept in the warmest environment so they will lose the least heat *[1 mark]*. Restricting movement and heat loss saves energy, so the transfer of energy from the feed to the turkeys is more efficient *[1 mark]*.
 b) E.g. the turkey meat is cheaper to buy *[1 mark]*.

Page 98: Problems with Food Production and Distribution

1 a) E.g. factory farmed eggs are cheaper for the customer *[1 mark]* and may allow the farmers producing the eggs to have a better standard of living (as they make more money) *[1 mark]*. The production of factory farmed eggs also helps to feed an increasing population compared to free range eggs, as more food is produced more quickly *[1 mark]*. However, many people think that keeping birds in factory farm conditions is cruel *[1 mark]*. The crowded conditions create a favourable environment for the spread of disease among the birds *[1 mark]*. Also, antibiotics are more likely to be used in factory farmed birds, which means that antibiotics can become less effective as human medicines *[1 mark]*. Energy is used to keep factory farmed birds warm — this means that more fossil fuels may be burnt to produce factory farmed eggs compared to free range eggs *[1 mark]*. *[Maximum of 4 marks.]*
 b) Food that has travelled further/has more 'food miles' will have used more fuel to transport it *[1 mark]*. This releases more carbon dioxide into the atmosphere and so contributes to global warming *[1 mark]*.
2 a) E.g. so that there will still be enough fish to eat in the future / to stop species of fish from disappearing in some areas *[1 mark]*.
 b) E.g. introduce fishing quotas *[1 mark]* to restrict how many fish can be taken from an area, allowing stock to recover *[1 mark]*. Use nets with larger holes *[1 mark]*, so smaller/younger fish/fish of the wrong species are not caught *[1 mark]*.

Pages 99-100: Decay and the Carbon Cycle

1 a) i) photosynthesis *[1 mark]*
  ii) carbon dioxide *[1 mark]*
 b) i) respiration *[1 mark]*
  ii) Plants convert carbon (from carbon dioxide) during photosynthesis into glucose/carbohydrates/fats/other substances in their cells *[1 mark]*. When animals eat plants, the carbon contained in these substances is transferred to the animals *[1 mark]*.
 c) i) Fossil fuels are formed from dead animals and/or plants, which contain carbon *[1 mark]*.
  ii) Carbon is released into the atmosphere as carbon dioxide when fossil fuels are burnt *[1 mark]*.
2 a) Warm, moist and aerobic conditions. *[2 marks for all three conditions, 1 mark for at least two conditions]*.
 b) carbon dioxide *[1 mark]*

156

3 a)   A stable community is one in which materials removed from the environment/soil are balanced by those that are put back in *[1 mark]*.

b)   How to grade your answer:

0 marks: There is no relevant information.

1-2 marks: There is a brief explanation of how materials are cycled through the lawn.

3-4 marks: There is some explanation of how materials are cycled through the lawn.
The answer has a logical structure and spelling, punctuation and grammar are mostly correct.

5-6 marks: There is a clear and detailed explanation of how materials are cycled through the lawn to create a stable community. The answer has a logical structure and uses correct spelling, grammar and punctuation.

Here are some points your answer may include:

The grass in the lawn absorbs elements such as nitrogen from the soil and carbon dioxide from the air.

It uses these chemicals to make complex compounds (including proteins, carbohydrates and fats) and for growth.

When the lawn is cut and the cuttings are left on the grass, the cuttings decay as microorganisms break them down/ digest them.

This decay process releases nitrogen and other substances back into the soil.

As the microorganisms digest the grass, they respire, releasing carbon dioxide back into the air.

The nitrogen and carbon dioxide are then available to be absorbed by the grass again.

The materials that are removed from the environment are balanced by the materials that are put back in, which creates a stable community.